Advanced Filters and Components For Electrical Power Applications

Prepared By

Timothy C. Neugebauer
Brandon J. Pierquet
David J. Perreault

Laboratory for Electromagnetic
And Electronic Systems

Massachusetts Institute of Technology

Wexford Press
2008

Contents

Chapter 1

Introduction

1.1 Background

No real electrical element is ideal. Wherever a current loop exists, an inductance can be found. Wherever two conductors are near each other, a capacitance can be found. Real electrical elements can be well modeled by their ideal element for only the range of frequencies in which the parasitic components are negligible. Outside of this range the impedance of real electrical elements will be dominated by their parasitic components. To reflect this, the real electrical element can be modeled with a collection of ideal parasitic inductors, capacitors, and resistors. These models classify the parasitics that influence the high- and low-frequency operation of real elements. This report addresses the design of filter and filter components that compensate for the parasitic effects present in electrical elements, enabling dramatic improvements in filter performance.

1.1.1 Capacitors

High frequency models for capacitors have been well established [1]. One such model, shown in Fig. 1.1a, shows an ideal capacitor in series with a resistor (R_{ESR}) and an inductor (L_{ESL}). This parasitic resistance is a combination of the lead resistance and the dielectric losses in the capacitor. (The dielectric resistance is in parallel with the capacitor but it is usually lumped into a frequency dependant series resistance.) The parasitic inductance is caused by the capacitor current and its associated stored magnetic energy storage associated with the current loop. At low frequencies the impedance of the capacitor is dominated by the ideal capacitance in the model. (At very low frequencies the capacitor impedance is dominated by a contact surface resistance or insulation resistance that is in parallel with the capacitance; this parasitic is not considered here.) At higher frequencies though, the inductive parasitic elements will dominate the performance of the capacitor. Figure 1.1b schematically shows a typical capacitor's impedance as a function of frequency. Figure 1.1c shows an impedance plot of an X-type (safety) capacitor (Beyschlag Centrallab 2222 338 24 224, 0.22 μF, 275 Vac). The frequency in which a capacitor changes from acting capacitive to inductive is called the self-resonant frequency. For example, a typical aluminum electrolytic capacitor may appear inductive (impedance rising with frequency) at frequencies above 50-100 kHz, limiting its ability to shunt ripple at high frequencies. Similarly, large-valued film capacitors typically become inductive in the range of 100 kHz - 1 MHz. While not perfect, this capacitor model forms a good basis for developing improved components and filters.

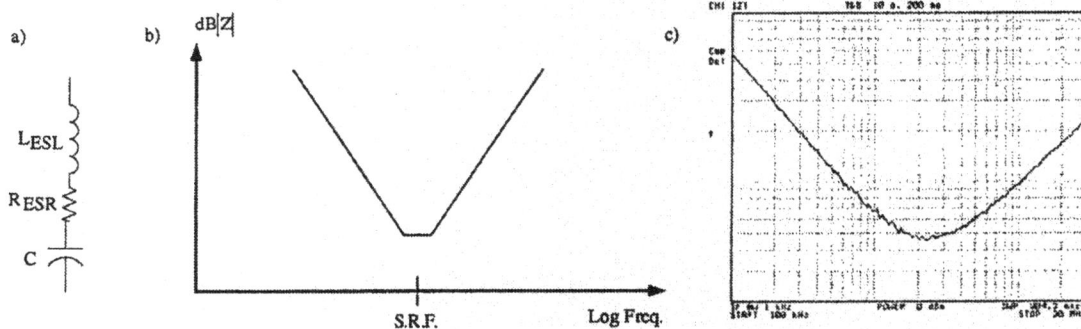

Figure 1.1: (a) The high frequency model of a typical capacitor, (b) the magnitude of the impedance as a function of frequency and (c) the plot of the magnitude of the impedance of an X-type (safety) capacitor (Beyschlag Centrallab 2222 338 24 224, 0.22 μF, 275 Vac).

1.1.2 Inductors

Inductors can also be modeled with two main parasitic components as shown in Fig. 1.2a. There exists a parasitic resistance in series with an ideal inductor and a parasitic capacitance in parallel with the inductor. This resistor models the power losses in an inductor including the winding resistance of the wire and magnetic losses in the core. (The magnetic losses represent a resistance in parallel with the inductor; this resistance is usually converted into a frequency dependent series resistance. In general, the effective resistance that is observed is frequency dependent, but we neglect this effect at present.) The parasitic capacitance exists because of the proximity of the windings to each other. Any two individual turns will have different voltages (due to induction and voltage drops caused by the winding resistance), and the displacement currents that consequently flows across turns can be modeled with a parasitic capacitance. This parasitic capacitance depends on the size and arrangement of the windings. The parasitic resistance affects the impedance of the inductor at low frequencies. At high frequencies the parasitic capacitance will dominate the impedance of the inductor. Figure 1.2b schematically shows the impedance of the inductor across a wide range of frequencies. Figure 1.2c shows the impedance of the Schott 67127100 680μH inductor. The self-resonant frequency of inductors are often significantly greater than the self-resonant frequency of capacitors for many filter designs of practical interest. The self-resonant frequency of a power inductor is typically in the range of 5-10 MHz and higher though in common-mode inductor designs it is often lower. Again, the model of Fig. 1.2b is a simplification of the actual behavior of practical devices, but it is a good basis for developing improved components and filters.

1.1.3 Filters

Electrical filters are designed to prevent unwanted frequency components from propagating from the filter input port to the filter output port, while passing desirable components. In power applications, filters are important for attenuating electrical ripple, eliminating electromagnetic interference (EMI) and susceptibility, improving power quality, and minimizing electromagnetic signature. One byproduct of switching power converters is the generation of unwanted high frequency signals at the input and output of the converter. EMI filters for DC outputs typically employ capacitors as

5

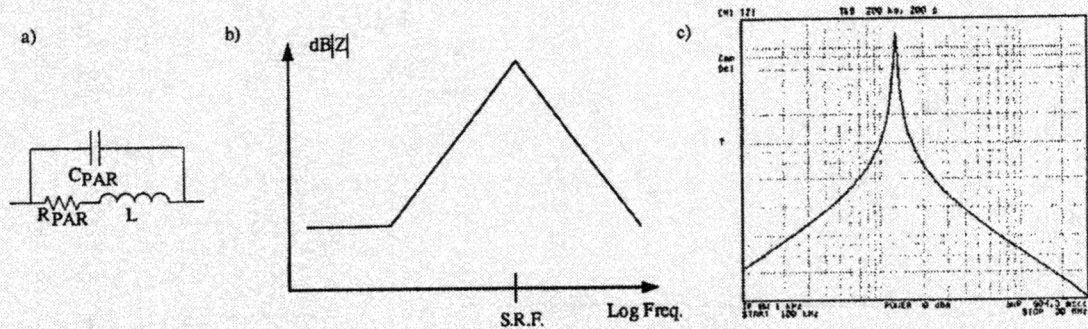

Figure 1.2: (a) The high frequency model of a typical inductor, (b) the magnitude of the impedance as a function of frequency and (c) the plot of the magnitude of the impedance for an inductor (Scohtt 67127100 680/muH.)

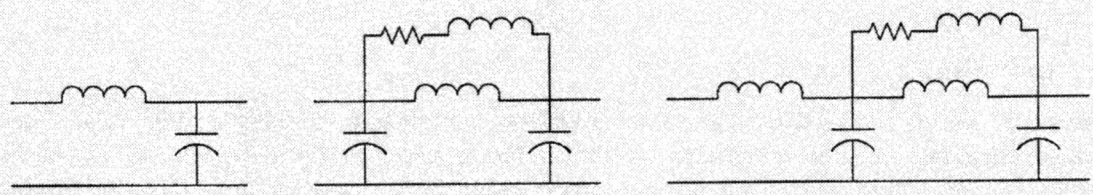

Figure 1.3: Some common low-pass filter structures for power applications.

shunt elements, and may include inductors as series elements, as illustrated in Fig. 1.3.

The attenuation of a filter stage is determined by the amount of impedance mismatch between the series and shunt paths. For a low-pass filter, minimizing shunt-path impedance and maximizing series-path impedance at high frequencies is an important design goal. Design methods for such filters are described in [2] and [3], for example. The parasitic components of the capacitor and inductor will affect the performance of a filter. An ideal inductor and capacitor would be perfect for the series and shunt paths respectively. However, an ideal low pass filter cannot be made with non-ideal components. The parasitic elements in the filter will, at some frequencies, control the attenuation of the filter.

Let us examine the operation of a two element filter with a resistive load that is shown in Fig. 1.4a with all of its high frequency parasitic elements. A simulation of the filter is shown in Fig. 1.4b. There are three different frequency ranges to examine. At frequencies lower than the self-resonant frequency of the capacitor, the filter behaves like an ideal low pass filter. The mid-frequency range is above the self-resonance of the capacitor and below the self-resonance of the inductor. In this range the filter operates like a voltage divider made up of inductors. The attenuation is constant and approximately

$$\frac{L_{ESL} \parallel L_{\text{Load}}}{L_{ESL} \parallel L_{\text{Load}} + L} \tag{1.1}$$

Where L_{Load} is the parasitic inductance of the load resistor. Note that in order to increase performance in the mid-frequency range increasing the capacitance is not helpful. Typically, larger

6

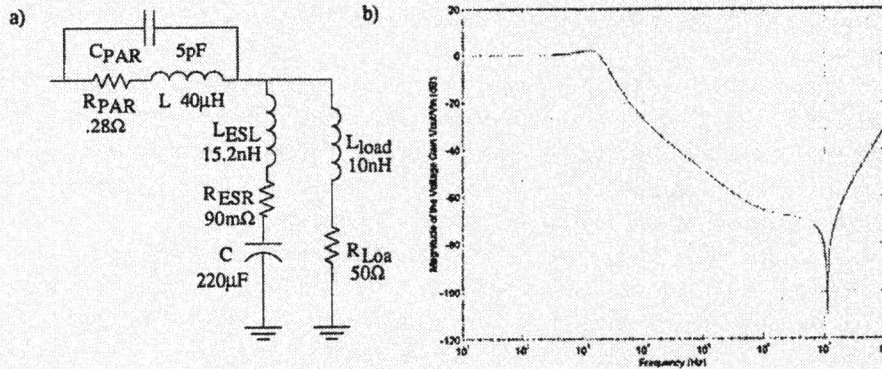

Figure 1.4: (a) A low pass filter made up of high frequency component models and (b) the voltage gain of the filter based on frequency. At frequencies below 10kHz the low pass filter looks ideal, but after 1 MHz the gain is not decreasing by 40 dB per decade and is relatively constant and after 10 MHz the gain increases.

capacitors have more equivalent series inductance which, for this frequency range, lowers the amount of attenuation. The traditional approach to overcoming filter capacitor limitations is to parallel capacitors of different types (to cover different frequency ranges) and/or to increase the order of the filter used (i.e., using more filter elements in a larger filter design). Both of these approaches can add considerable size and cost to the filter.

At frequencies above the self-resonance of the inductor, the capacitor is dominated by its parasitic series inductance and the inductor is dominated by its parasitic parallel capacitance, thus the capacitor looks inductive and the inductor looks capacitive. In this range the filter operates like a high pass filter, with an attenuation of

$$\frac{V_{out}}{V_{in}}(s) = \frac{(L_{ESL} \parallel L_{\text{Load}})C_{par}s^2}{(L_{ESL} \parallel L_{\text{Load}})C_{par}s^2 + 1} \tag{1.2}$$

At high enough frequencies, additional parasitics that have not been modeled will affect the performance of the filter. These higher order parasitic elements are usually undefined since their behavior is difficult to predict and the presence of the modeled parasitic elements usually dominate the performance of the component over the frequency range of interest.

1.2 Research Objectives and Motivation

The objective of the research in this report is to improve the high frequency performance of components and filters by better compensating the parasitic effects of practical components. The main application for this improvement is in design of low pass filters for power electronics, although some other applications will be presented.

In power electronics the input and output filters are a dominant consideration in electromagnetic compatibility and often represent a major contribution to the weight, volume and cost of the system. Therefore, aspects of the design of the system, especially those related to EMI, are limited by the high frequency performance of the filters. The usual methods of improving the high frequency performance of the filter includes using more filter components. Filter performance can improve by

7

using more filter components and filter stages and higher quality inductors and capacitors. These methods add significant cost to the design of the system.

If the effect of high-frequency parasitic elements in the components can be reduced (at a low cost) the performance of the filter can be enhanced. This allows the development of filters with much better high frequency attenuation, or the reduction of filter size and cost at a constant performance level. In filtering and other applications, the ability to reduce the effect of parasitic elements will be a technique that will enable many high-frequency designs.

Specifically, this report will present two techniques that can be used to reduce the effects of parasitic inductance and capacitance. One technique, called inductance cancellation, is used to reduce the effect of parasitic inductance in a path of interest. The other technique, capacitance cancellation, will reduce the effect of a parasitic capacitance in an inductor. The techniques introduced here cannot be used to improve performance of passive components in all applications. These techniques, though, do provide major improvements in most filtering applications, in which parasitic components play a central role.

1.3   Report Overview

This report will introduce inductance cancellation and capacitance cancellation techniques. These techniques and their effects will be demonstrated in a wide range of circuits. Design guidelines for practical application of these principles are also developed and experimentally validated.

Chapter 2 fully explains the principles of inductance cancellation and experimentally demonstrates the use of the approach. Chapter 3 examines the design of filters with inductance cancellation in which the circuitry realizing cancellation is implemented using traces on a printed circuit board (pcb). Generally, the area needed on the pcb is comparable to the footprint of the component and can be placed under the component. Chapter 4 examines the design of components in which inductance cancellation windings are integrated with a capacitor to form an integrated filter element. Chapter 5 examines a design approach in which inductance cancellation can be adjusted with active control, allowing its use in applications where the parasitic inductance is not well known or controlled. Chapter 6 introduces cancellation for multiple capacitors, for use in common- and differential-mode filters. Chapter 7 looks at some uses of inductance cancellations in applications other than filtering. Chapter 8 introduces the capacitance cancellation technique and develops its application for improving the performance of inductors and common-mode chokes. Finally, Chapter 9 concludes the report.

Chapter 2

Filters and Components with Inductance Cancellation

2.1 Introduction

Inductance cancellation is a passive circuit technique that effectively shifts inductance from a circuit branch where it is undesirable to other branches where it is acceptable. In circuit terms, a consequence of the technique to be proposed will essentially provides a negative inductance in one circuit branch, and larger positive inductances in other circuit branches. The negative inductance can be placed in series with an unwanted parasitic inductance, thereby improving the high-frequency performance of the circuit. Therefore, the total inductance in the system will increase, not decrease, when inductances in one branch is shifted to two other branches.

The technique of inductance cancellation is well suited to improve the performance of capacitors, especially for their use in electrical filters. Capacitors are critical elements in such filters, and filter performance is strongly influenced by the capacitor's parasitics. This chapter introduces the application of this new design technique to overcome the capacitor parasitic inductance that limits filter performance at high frequencies. Coupled magnetic windings are employed to effectively cancel the parasitic inductance of capacitors, while adding inductance in filter branches where it is desired. The underlying basis of the new technique is treated in detail, and its application to the design of both discrete filters and integrated L-section filter components is described. Numerous experimental results demonstrating the high performance of the approach in both discrete filters and integrated filter elements are provided.

2.2 Inductance Cancellation

2.2.1 End-tapped Transformers

Magnetically-coupled windings can be used to cancel the effects of capacitor parasitics. Fig. 2.1 illustrates one possible connection of coupled magnetic windings, which we hereafter refer to as an "end-tapped" connection. In this case, each winding links flux with itself and mutually with the other winding. An electromagnetic analysis of this system leads to an inductance matrix description:

9

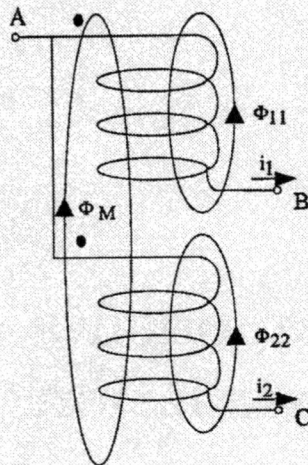

Figure 2.1: An end-tapped connection of coupled magnetic windings.

$$
\begin{bmatrix} \lambda_1 \\ \lambda_2 \end{bmatrix} = \begin{bmatrix} \frac{N_1^2}{\Re_{11}} + \frac{N_1^2}{\Re_M} & \frac{N_1 N_2}{\Re_M} \\ \frac{N_1 N_2}{\Re_M} & \frac{N_2^2}{\Re_{22}} + \frac{N_2^2}{\Re_M} \end{bmatrix} \begin{bmatrix} i_1 \\ i_2 \end{bmatrix} = \begin{bmatrix} L_{11} & L_M \\ L_M & L_{22} \end{bmatrix} \begin{bmatrix} i_1 \\ i_2 \end{bmatrix} \tag{2.1}
$$

where the flux linkages λ_1 and λ_2 are the time integrals of the individual coil voltages, and i_1 and i_2 are the individual coil currents. The self-inductances L_{11} and L_{22} and mutual inductance L_M are functions of the numbers of coil turns and the reluctances $\Re_{11}$, $\Re_{22}$, and $\Re_M$ of the self and mutual magnetic flux paths. In cases where no magnetic material is present, the behavior of the coupled windings is determined principally by the geometry of the windings. Conservation of energy considerations require that the mutual coupling between the windings be less than or equal to the geometric mean of the self-inductances. That is,

$$
L_M \leq \sqrt{L_{11} L_{22}} \tag{2.2}
$$

Thus, the inductance matrix of (2.1) is necessarily positive semidefinite. Note that while the constraint (2.2) limits L_M to be less than or equal to the geometric mean of L_{11} and L_{22}, it may still be larger than one of the two inductances. For example, with proper winding of the coils one may have

$$
L_{11} < L_M < \sqrt{L_{11} L_{22}} < L_{22} \tag{2.3}
$$

Figure 2.2 shows one possible equivalent circuit model for the coupled inductor windings based on the inductance matrix of (2.1). This model is referred to as the "T" model of the coupled windings and is derived in Appendix A. With the ordering of self and mutual inductances of (2.3), the inductance of one leg of the T model - the vertical leg in Fig. 2.2 - is clearly negative! It is this "negative inductance" that will be used to overcome the high-frequency limitations of filter capacitors.

The negative inductance effect arises from electromagnetic induction between the coupled windings. This is readily seen in the physically-based circuit model of the coupled windings shown in

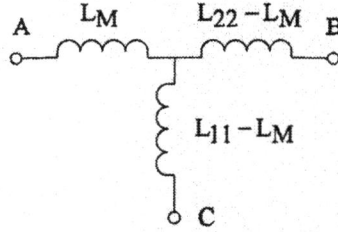

Figure 2.2: An equivalent circuit model for end-tapped coupling magnetic windings.

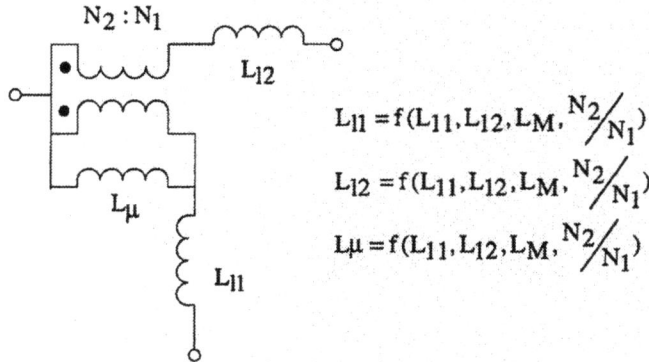

$$L_{11} = f(L_{11}, L_{12}, L_M, {}^{N_2}\!/_{N_1})$$

$$L_{12} = f(L_{11}, L_{12}, L_M, {}^{N_2}\!/_{N_1})$$

$$L_\mu = f(L_{11}, L_{12}, L_M, {}^{N_2}\!/_{N_1})$$

Figure 2.3: A physically based circuit model of the coupled magnetic windings. The formulae converting from these parameters to L_{11}, L_{22}, and L_M are in Appendix A.

Fig. 2.3. (With appropriate parameter values, the circuit models of Fig. 2.2 and Fig. 2.3 have identical terminal characteristics, and each captures the behavior of the system (2.1).) Appendix A examines several popular transformer models and demonstrates the conversion of the physically based model of Fig. 2.3 to the form shown in Fig. 2.2, which is the most useful representation of the transformer in this application. We stress that the negative inductance in the T model does not violate any physical laws. Only one leg of the T model has a negative inductance. The total inductance seen across any winding is - as expected - the positive-valued self inductance of the winding.

Fig. 2.4 shows the application of the coupled magnetic windings to a capacitor whose equivalent series inductance (ESL) is to be cancelled. The coupled windings are modeled with the T network of Fig. 2.2, while the capacitor is shown as an ideal capacitor C in series with parasitic resistance R_{ESR} and parasitic inductance L_{ESL}. (We also lump any interconnect parasitics into these elements.) When $L_{11} - L_M$ is chosen to be negative and close in magnitude to L_{ESL}, a net shunt path inductance $\Delta L = L_{11} - L_M + L_{ESL} \approx 0$ results. The combined network is very advantageous as a filter. A near-zero shunt path impedance (limited only by ESR) is maintained out to much higher frequencies than is possible with the capacitor alone. Furthermore, as L_{22} is much greater than L_M, the series-path inductance $L_{22} - L_M$ serves to either increase the order of the filter network or is in series with another filter inductor, both options will improving filter performance.

The voltage stress across the capacitor will go up slightly. Assume that, at a particular fre-

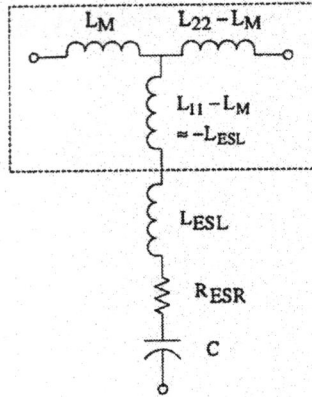

Figure 2.4: Application of coupled magnetic windings to cancel the series inductance of a capacitor. Capacitor ESR and ESL are shown explicitly, along with the equivalent T model of the magnetic windings.

quency, 99 % of the ac current will travel through the capacitor rather than the load. Also assume that using inductance cancellation will improve filtering by a factor of 10. Therefore with inductance cancellation 99.9 % of the ac current will travel through the capacitor. The increase in ac current will correspond to a small increase in voltage across the capacitor.

An improvement in filter performance can be seen by examining the effect of inductance cancellation on (1.1) which is repeated below.

$$\frac{L_{ESL} \parallel L_{\text{Load}}}{L_{ESL} \parallel L_{\text{Load}} + L} \tag{2.4}$$

This equation gives the approximate attenuation of a second order low pass filter in a frequency range in which the parasitic inductance dominates the impedance of the capacitor. Two things should be considered when examining this equation when inductance cancellation is applied. Normally to improve the attenuation the filter needs either a larger inductor or a better quality capacitor (with less parasitic inductance). With inductance cancellation the value of L_{ESL} will be reduced by a factor of 10 or more and the values of L_{Load} and L will be increased by an amount larger than L_{ESL} but on the same order of magnitude. The term $L_{ESL} \parallel L_{\text{Load}}$ will thus be greatly reduced by inductance cancellation and improve the filter performance. Also, because of inductance cancellation the frequency at which parasitic inductance starts to dominate the performance of the capacitor will be higher, and therefore the filter will look like a second order system for a wider range of frequencies.

In the previous chapter it was stated that all inductors (and transformers) have parasitic capacitors that will impair their performance at high frequency. The inductance cancellation transformer should be designed so as to have a negligible capacitance. Typical inductors and transformers with magnetic cores have a limited winding area and a large desired inductance, in order to minimize the volume of the structure, the windings are packed in close proximity to each other. Transformers for inductance cancellation do not need a lot of turns to achieve the desired inductances and are not limited to a prescribed winding area. Also, the self-resonant frequency of the transformer will be extremely high (typically far higher than the conduction EMI frequency range) since the inductances are so small.

12

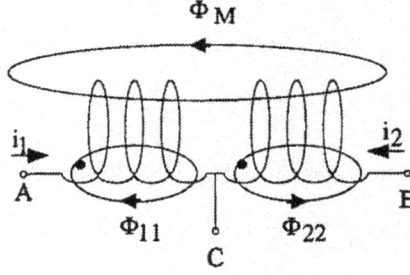

Figure 2.5: A center-tapped coupled magnetic winding configuration.

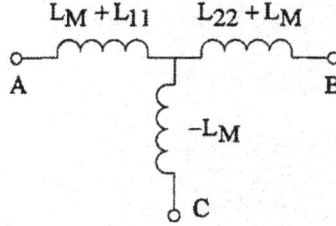

Figure 2.6: An Equivalent circuit model for the center-tapped coupled magnetic windings.

2.2.2 Center-tapped Transformers

It should be appreciated that the other connection method of the magnetic winding structure can also be used to realize inductance cancellation. Another three-terminal coupled magnetic structure that can be used is shown in Fig. 2.5. This implementation is advantageous in that it can be formed from a single winding tapped at an appropriate point. An electromagnetic analysis of the system of Fig. 2.5 results in an inductance matrix:

$$
\begin{bmatrix} \lambda_1 \\ \lambda_2 \end{bmatrix} = \begin{bmatrix} \frac{N_1^2}{\Re_{11}} + \frac{N_1^2}{\Re_M} & \frac{-N_1 N_2}{\Re_M} \\ \frac{-N_1 N_2}{\Re_M} & \frac{N_2^2}{\Re_{22}} + \frac{N_2^2}{\Re_M} \end{bmatrix} \begin{bmatrix} i_1 \\ i_2 \end{bmatrix} = \begin{bmatrix} L_{11} & -L_M \\ -L_M & L_{22} \end{bmatrix} \begin{bmatrix} i_1 \\ i_2 \end{bmatrix} \tag{2.5}
$$

where the self inductances L_{11} and L_{22} and mutual inductance L_M are again functions of the numbers of coil turns N_1, N_2 and the reluctances of the respective magnetic flux paths. The magnitude of the mutual inductance is again limited by the constraint (2.2), though without the ordering imposed in (2.3).

The terminal characteristics of the system of Fig. 2.5 can be modeled with the "T model" of Fig. 2.6 following the steps outlined in Appendix A.3. Again, one branch of the T model has a negative inductance (in this case equal in magnitude to the mutual inductance L_M). When L_M is chosen to be close in magnitude to the equivalent series inductance L_{ESL} of an electrical circuit path (e.g., through a capacitor) connected to the bottom terminal, a reduced net effective inductance $\Delta L = -L_M + L_{ESL} \approx 0$ results in the capacitor's shunt path.

As described above, coupled magnetic windings are used to cancel inductance in the capacitor branch (e.g., due to capacitor and interconnect parasitics) and provide filter inductances in the other branches. In a low-pass filter, this corresponds to a cancellation of the filter shunt branch

13

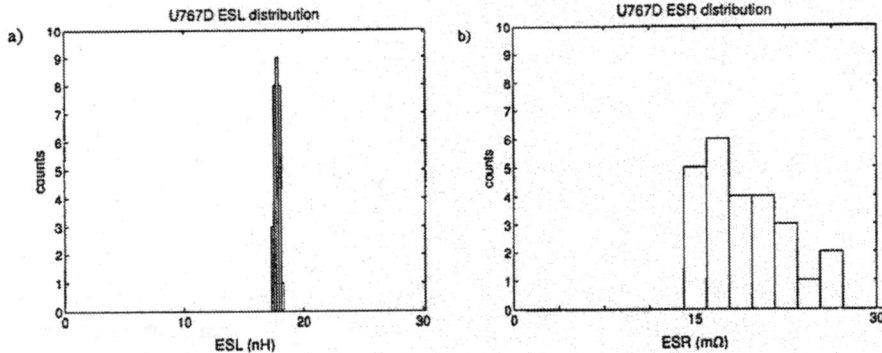

Figure 2.7: (a) ESL and (b) ESR histograms for 30 United Chemi-Con U767D 2200 μF 35 V capacitors. ESL range: 17.29 to 18.13 nH, $\sigma = 44.6$ pH. ESR range: 14.2 $m\Omega$ to 60.9 $m\Omega$ (outlier not shown).

inductance, and an addition of series branch inductance. (The final branch path necessarily has an inductance greater than or equal to the magnitude of the "negative" inductance that is introduced in the capacitor path.) We point out that the use of coupled magnetics in filters is not in itself new. In fact, use of coupled magnetic windings in filters dates at least as far back as the 1920's [4], and has continued up to the present time in many forms [5–10] (see [5] for a good review of such usage). The approach described here differs from these existing methods in that the coupling of the windings is used to cancel the effects of parasitic inductance in the capacitor and interconnects, permitting dramatic improvements in filtering performance to be achieved.

2.3 Implementation

In this section we consider application of this inductance cancellation technique to the design of both discrete filters and integrated filter components. One important design consideration is that of variability: if tuning of individual units is to be avoided, the inductances of both the capacitor and the magnetic windings must be consistent from component to component[1]. Fortunately, unlike capacitance or ESR values, capacitor ESL is typically consistent to within a few percent. For example, the histograms of Figs. 2.7(a) and 2.7(b) show the distribution of ESL and ESR for a type of electrolytic capacitor that is widely used in filters. The ESR varies over a wide range from 14.2 $m\Omega$ to 60.9 $m\Omega$ (outlier not shown). The ESL, by contrast, varies only from 17.29 nH to 18.13 nH (with a standard deviation of 44.6 pH), representing a maximum variation in ESL of only $\pm 2.4\%$ across units. This makes sense: the absence of magnetic materials means that the inductance of the structure depends primarily on geometry, while capacitance and resistance depend on material and interface properties. One may conclude that inasmuch as appropriate coupled-magnetic structures can be created, the parasitic inductance can be repeatably cancelled to within a few percent of its original value.

The capacitor inductance to be cancelled in a practical design is typically quite small (on the order of tens of nanohenries). Coupled magnetic windings appropriate to the cancellation technique

---

[1]Systems incorporating active self tuning [11–13] (e.g., via controllable magnetics [14–17]) are also possible. We defer consideration of this approach to chapter 5.

must thus be able to accurately generate a negative effective shunt inductance in this range under all operating conditions. One approach for achieving this is to use coupled windings without magnetic materials. Such "air-core" magnetics are appropriate given the small inductances needed and the desire for repeatability and insensitivity to operating conditions.

Two approaches for employing the proposed inductance cancellation technique are considered in this report. We first address the use of inductance cancellation methods in the design of filter circuits built with discrete components (e.g., capacitors and inductors) using conventional manufacturing techniques. We then explore the integration of cancellation windings with a capacitor to form an integrated filter element - a three terminal device providing both a shunt capacitance (with extremely low shunt inductance) and a series inductance.

2.3.1 Discrete Filters

An immediate application of the proposed technology is in the design of discrete filters - that is, filters built with available or easily manufactured components using conventional fabrication techniques. In this approach, a coupled winding circuit is connected to a discrete capacitor to provide a very low-inductance path through the capacitor along with a second high-inductance path. The coupled winding circuit should have repeatable inductance parameters (that are properly matched to the capacitor), and should have minimal size and cost impact on the filter.

One simple implementation method is to print the coupled windings as part of the filter printed circuit board (PCB). Printing the magnetic windings on the PCB results in extremely repeatable magnetic structures and interconnects. Furthermore, it represents essentially no extra cost or volume in the design if the PCB space underneath the filter capacitor can be used for the windings.

We have found air-core PCB windings to be highly effective for the proposed inductance cancellation technique. As will be demonstrated in Section 2.4, practical printed PCB windings can be implemented using either end-tapped (Fig. 2.1) or center-tapped (Fig. 2.5) winding configurations, and can be placed either partially or entirely underneath the capacitor on the PCB. A two-layer circuit board is typically sufficient to implement the windings with the required interconnects accessed at the outside of the spiral windings. The coupled winding circuits demonstrated here were designed using a widely-available inductance calculation tool [18] and refined experimentally.

2.3.2 Integrated Filter Elements

In addition to their application in discrete filters, inductance cancellation techniques have application to new filter components. Here we introduce the integration of coupled magnetic windings (providing inductance-cancellation) with a capacitor to form an integrated filter element - a single three-terminal device providing both a shunt capacitance (with extremely low inductance) and a series inductance. To do this, one can wind inductance-cancellation magnetics on, within, or as part of the capacitor itself. This approach, illustrated in Fig. 2.8, minimizes the volume of the whole structure, as the same volume is used for the capacitive and magnetic energy storage. For example, starting with a wound (tubular) capacitor, one could wind the coupled magnetics directly on top of the capacitor winding. The magnetic windings can also be implemented through extension or patterning of the capacitor foil or metallization itself. An integrated filter element utilizing inductance cancellation may be expected to have far better filtering performance than a capacitor of similar size. We note that components incorporating both capacitive and inductive coupling have a long history in power applications [19–25] and continue to be an important topic of

15

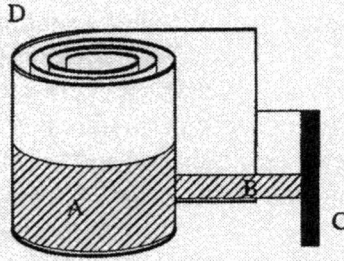

Figure 2.8: Integrated filter element D is constructed by adding magnetically-coupled windings A and B over, or as part of, the basic capacitor structure. The integrated filter element is then a three-terminal device, with the connection of the two magnetic windings brought out as terminal C.

research (e.g., [26–29]). However, the aims and resulting characteristics of such prior art integrated elements are quite different than those described here. The approach described here is different in that magnetically-coupled windings are used to nullify the effects of the parasitic inductance in the capacitive path. This permits, with relatively modest changes in manufacturing methods, dramatic improvements in filtering performance to be achieved as compared to conventional components.

As with discrete filters, both end-tapped and center-tapped coupled-winding configurations are possible. (Note that in some integrated implementations, flux associated with current flow in the capacitive element may link the cancellation windings. This changes the details of the magnetic analysis - and may be used to advantage - but the underlying principles remain the same.) Consider an integrated component having a wound structure, as suggested by Fig. 2.8. In an end-tapped configuration, the magnetic windings comprise two conductors co-wound and electrically connected at one end (one terminal of the three terminal device.) The other end of one conductor is a second terminal of the device. The other end of the second conductor is connected to one plate of the capacitor. (The magnetic winding may be formed as a direct continuation of the capacitor winding in this case.) The other plate of the capacitor is connected to the third terminal. A basic magnetic analysis of this structure will assume that there are no fringing fields and flux due to parasitic inductances are well shielded by the capacitor, therefore the capacitor can be modeled as a cylindrical metal enclosure. A more advanced version will model several of the outer layer of the capacitor but the significance of the fringing fields will be low. Regardless of which model is used, a design iteration can be used to identify a design that has consistent performance.

In a center-tapped magnetic winding configuration, the coupled magnetic windings may be formed as a single conductor wound concentrically with the capacitor windings. The magnetic winding is tapped (connected to one plate of the capacitor) at a specified point in the winding. The other plate of the capacitor and the two ends of the magnetic winding form the three terminals of the device. Fig. 2.9 illustrates one possible method for forming the cancellation winding over the capacitor structure and interconnecting it to one capacitor plate.

Another manner of incorporating inductance cancellation directly into the capacitor is to build small transformers and to connect them to the leads inside the capacitor. A loop of metal can be made into a transformer, this transformer can be encapsulated in a non-conducting material. Three connection points can be made from the transformer to the rest of the capacitor. The capacitor packaging can enclose both the capacitor and the transformer.

16

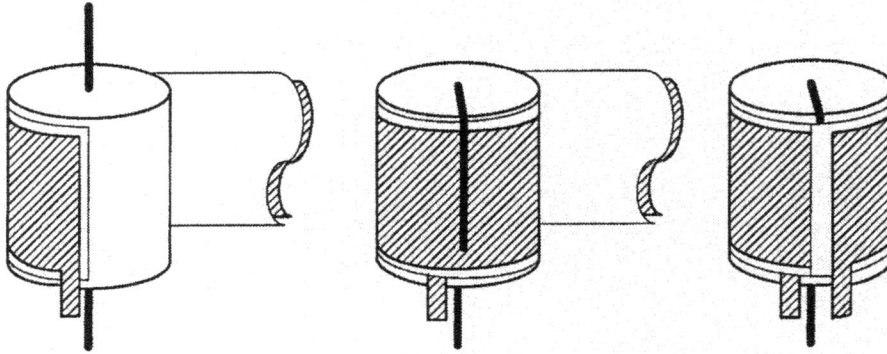

Figure 2.9: One construction method for an integrated filter element with a center-tapped winding.

2.4 Experimental Results

In this section we demonstrate the viability and high performance of the proposed inductance cancellation technology. We validate the approach for both discrete filters and integrated filter elements across a variety of capacitor sizes and types, and with both end-tapped and center-tapped winding configurations. The choice between end-tapped and center-tapped transformers were made arbitrarily in this section so as to show examples with each configuration. In the next chapters comparison between the two transformer styles will be made. We also demonstrate the large performance advantage of a prototype integrated filter element in a power converter application.

2.4.1 Evaluation Method

To evaluate the effectiveness of the inductance cancellation method, the test setup of Fig. 2.10 is used. The device under test (DUT) is either a capacitor, a capacitor plus PCB cancellation windings, or an integrated filter element. The DUT is driven from the 50 Ω output of the network analyzer. As the driving point impedance of the DUT is always far less than the output impedance of the network analyzer, the drive essentially appears as a current source. The voltage response is measured at the 50 Ω load of the network analyzer. The input impedance of the network analyzer is much greater than the impedance associated with the series output inductance of the DUT for the frequencies under consideration. Accordingly, this test effectively measures the shunt impedance of the DUT relative to the 50 Ω load impedance of the network analyzer. Thus, this test focuses on filtering improvements associated with the shunt-path inductance cancellation, while suppressing improvements available through the introduction of series path inductance. In the practical application of a filter, one could take advantage of the series inductance provided by the cancellation windings to further improve attenuation performance.

It should be noted that all measurements of capacitor performance at frequencies up to 30 MHz need to be carefully performed. A circuit layout that includes large parasitic inductive loops can induce a signal on par with (or greater than) the signal to be measured. To ensure proper measurement of filter performance the input and output connections to the network analyzer can be made with BNC to PCB connectors. Testing performed without these connectors used a pair of twisted wires to make the connection to the capacitor. In that setup the signals received by the

17

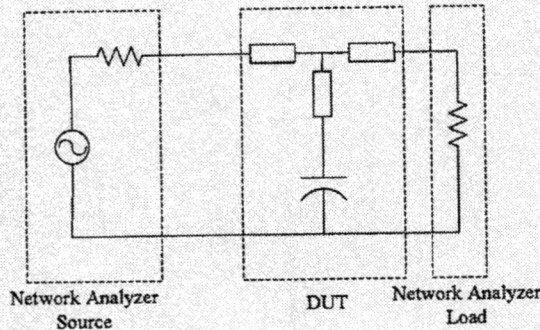

Figure 2.10: An Experimental setup for evaluating filters and components. The Network analyzer is an Agilent 4395A.

twisted pair were up to an order of magnitude greater than the output ripple to be measured. The resulting test setup had performance that was dependant on the positions of all the connections to the capacitor circuit. Further tests of the noise floor (a test in which the connection to the capacitor from the network analyzer was made with twisted pair, but the output is left disconnected) showed that the receiver would pick up significant amounts of radiated noise. All measurements made in a test stand with low inductance connections to the board did not have this problem.

2.4.2 End-Tapped Discrete Filter

Fig. 2.11 shows a discrete-filter implementation of the inductance cancellation technique for a large film capacitor (Cornell-Dubilier 935C4W10K, 10 μF, 400 V). The coupled magnetic windings are printed in the PCB with a rectangular and circular coil version on each board. The printed windings are placed in an end-tapped configuration, with the winding connected to the capacitor made from a single turn on the top side of the 0.031" thick board. The second winding is placed on the bottom side of the board and spirals inwards. A wire connects the output of the converter to a point on the spiral to maximize inductance cancellation in the system (approximately 2.5 turns are used, placing the output at the other end of the capacitor.) Current return for both the capacitor and filter output current is on a bottom-side ground plane adjacent to the capacitor. Note that the printed windings fit in the board space underneath the capacitor.

Fig. 2.12 shows comparative experimental results using the test arrangement of Fig. 2.10 for both a capacitor alone and a capacitor plus cancellation windings. The results are presented for frequencies up to 2 MHz, at 10 dB per division. The capacitor alone is self-resonant at a frequency of about 150 kHz, and acts as an inductor at higher frequencies. With cancellation windings, the effective shunt impedance of the filter drops to the value of the ESR and remains there to frequencies higher than 1 MHz. The unmodified capacitor will have a lower impedance in the area around the self resonant frequency because the impedance of the inductor and capacitor will have the same order of magnitude, but different signs.

At approximately 1.25 MHz the capacitor itself has a secondary resonance, above which its effective ESL and ESR change by several percent. The secondary resonance is not easily observed in the presented magnitude plot for the capacitor alone because it is overwhelmed by the primary ESL impedance. We have observed such secondary resonances in several large-valued film capacitors.

18

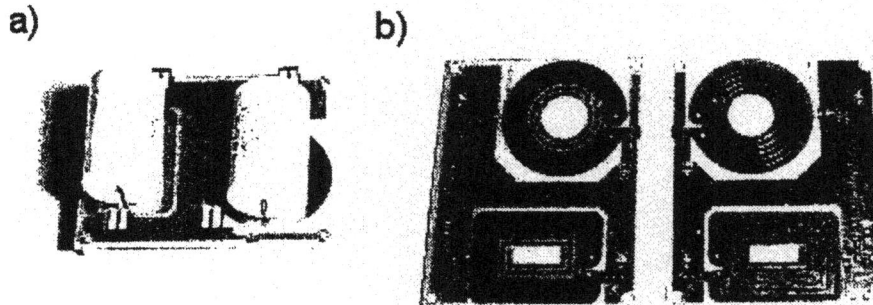

Figure 2.11: Discrete filters using Cornell Dubilier 935C4W10K capacitors with end-tapped cancellation windings printed in the PCB. The board in the middle shows the top (component) side of the board, while the board on the right shows the bottom side of the board.

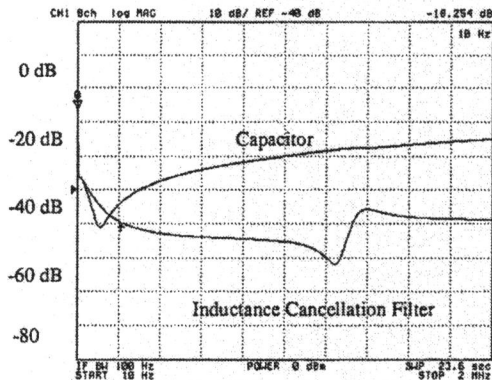

Figure 2.12: Performance comparison of a Cornell Dubilier 935C4W10K film capacitor and the inductance cancellation filter of Fig. 2.11.

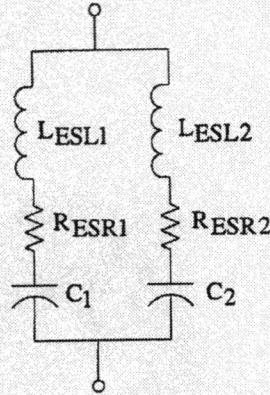

Figure 2.13: A high frequency model of a capacitor. The second resonance is set by C_2 and L_{ESL2}.

The secondary resonance can be described by looking at a distributed model for the capacitor. Let every layer of the capacitor be modeled as a capacitor with a series parasitic resistor and inductor and connect each of these layers in parallel. For the most part the capacitances, resistances and inductances are about equal and can easily be combined into a lump model, but the inner and outer layers of the capacitor will have a significant difference in values. Thus a better high frequency model of the capacitor is shown in Fig. 2.13. For typical capacitors, L_{ESL1} will dominate the high frequency operation of the capacitor and the secondary resonance will be negligible. Despite the higher frequency model parameters, the experimental results demonstrate that the inductance cancellation technique provides a large reduction in effective shunt-path inductance, resulting in a factor of 10 improvement (20 dB) in filtering performance out to very high frequencies. This dramatic improvement is achieved at no change in size or cost, since the cancellation windings fit in the board space beneath the capacitor.

2.4.3 Center-Tapped Discrete Filter

A discrete filter implementation based on an X-type (safety) capacitor (Beyschlag Centrallab 2222 338 24 224, 0.22 μF, 275 Vac) was also evaluated. Such capacitors are widely used in EMI filters for line applications. Inductance cancellation windings were again formed on the printed circuit board, this time using a center-tapped winding configuration. The coupled windings for each terminal comprised a single turn on the top layer of the board and a single turn on the bottom layer of the board. The traces are 100 mil wide and the single turn is a circle with an average radius of 462 mils. The test board is shown in Fig. 2.14

Fig. 2.15 shows comparative experimental results using the test setup of Fig. 2.10 for the X capacitor with and without the printed cancellation windings. The conducted EMI frequency range up to 30 MHz is considered, with results plotted at 10 dB per division. The capacitor alone has an ESR of approximately 45 $m\Omega$ and an ESL of approximately 10 nH, resulting in a self resonant frequency in the vicinity of 3 MHz. Addition of the cancellation windings results in dramatic improvement in filtering performance at frequencies above 5 MHz (as much as 26 dB). Careful impedance measurements suggest an equivalent T model with input and output branch inductances of approximately 17.5 nH each and a total shunt-branch inductance of approximately

Figure 2.14: Test board for measuring the performance of a BC 2222 338 24 224 X-capacitor using a center-tapped cancellation winding printed on the pcb.

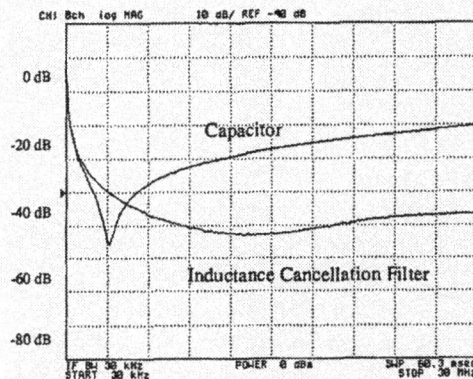

Figure 2.15: Comparison of a BC 2222 338 24 224 X-capacitor to a corresponding discrete filter with center-tapped cancellation windings.

1.2 nH, corresponding to an 88% reduction in effective capacitor-path inductance. The results confirm the dramatic filtering improvements possible with this technique.

2.4.4 Integrated Filter Element (Film)

We have also constructed prototype integrated filter elements by winding cancellation windings on the bodies of existing capacitors. Fig. 2.16 shows a prototype integrated filter element built with a center-tapped winding configuration using the construction method of Fig. 2.9. (The board also has a capacitor without cancellation windings.) The prototype filter element was constructed using a Rubycon MMW 106K2A film capacitor (10 μF, 100 V). The tapped cancellation winding was wound on the center of the capacitor body with 0.25" wide 1 mil thick copper tape, using 1 mil thick mylar tape for insulation. A 330° winding was placed on the capacitor body and tapped, followed by a continued 225° winding. Since this is an air core transformer, a transformers with partial turns is plausible. (With a magnetic core partial turns will lead to flux imbalances and a

21

Figure 2.16: A prototype integrated filter element based on a Rubycon MMW 106K2A film capacitor and center-tapped cancellation winding. The board used for comparison, consisting of only a normal capacitor, is also shown.

higher power loss). Partial turns can be formed can be placing the connections to the rest of the circuit anywhere around the filter element. (Thus, the inductance loop can be made by a partial turn and the board wiring) The prototype filter element was mounted on a two-sided pc board with the top side split between the filter input and output nodes, and a full-width ground plane on the bottom side of the board. (In the capacitor only case, the input and output nodes are the same.)

Test results using the experimental setup of Fig. 2.10 are shown in Fig. 2.17. The capacitor alone exhibits its primary self-resonance at approximately 600 kHz, with a second resonance in the vicinity of 3 MHz. (Again, this is characteristic of some capacitors.) The integrated filter element nullifies the principle ESL characteristic, making the effects of the secondary resonances more pronounced. Nevertheless, a tremendous improvement in filtration performance is obtained at high frequencies, exceeding 30 dB (a factor of 30) for frequencies above 7 MHz.

2.4.5 Integrated Filter Element (Electrolytic)

An integrated filter element based on a 2200 μF electrolytic capacitor (United Chemi-Con U767D, 2200 μF, 35 V) was also evaluated. Such capacitors are widely used in filters for automotive applications. The center-tapped cancellation winding was wound 0.25" from the top of the capacitor body with 0.5" wide 1 mil thick copper tape, using 1 mil thick mylar tape for insulation. A 3.25 turn winding was placed on the capacitor body and tapped, followed by a continued 3.125 turn winding. The test board setup is shown in Fig. 2.18

Test results using the experimental setup of Fig. 2.10 are shown in Fig. 2.19. The impedance of the capacitor alone begins to rise in the vicinity of 100 kHz due to ESL effects. By contrast, the integrated filter element continues to attenuate the input out to much higher frequencies, resulting in more than a 10 dB improvement at 200 kHz, and increasing to more than 20 dB above 1 MHz.

22

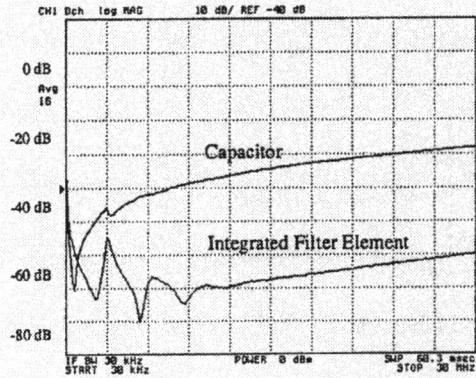

Figure 2.17: Comparison of a Rubycon MMW 106K2A film capacitor to the prototype integrated element of Fig. 2.16.

Figure 2.18: A prototype integrated filter element based on a United Chemi-Con U767D, 2200 μF, 35 V electrolytic capacitor and center-tapped cancellation winding. The board used for comparison, consisting of only a normal capacitor, is also shown.

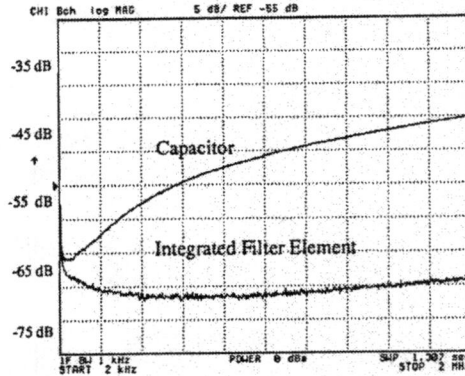

Figure 2.19: Comparison of a United Chemi-con U767D, 2200 μF electrolytic capacitor to the corresponding prototype integrated filter element.

2.5 Side Effects

The proposed techniques of inductance cancellation cannot be used arbitrarily to remove all parasitic inductances in a circuit. The main hindrances to this technique are the two additional positive inductances that exist in the T-model. In certain applications, such as EMI filtering, these extra inductances are beneficial, in some applications they can be ignored, but in many applications they will be detrimental.

For example, consider the capacitors used to maintain a constant voltage on a dc bus. The parasitic inductance of the capacitor does affect the power supply rejection ratio and noise on the bus. If inductance cancellation is used on a capacitor in this application, the two positive inductances in the T-model will cause a problem. When there is a surge of current on the bus, there will be an ac voltage across the inductor in series with the load. This ac ripple voltage will cause fluctuations in the bus voltage.

Another specification that needs to be considered is the amount of dc current in the filter element. Capacitors are a two terminal element and typically only the ac ripple current in the element is important. The capacitor with inductance cancellation is a two port filter element in which the dc current will flow through the air core transformer. Thus the windings of the transformer must be sized to carry the current. Problems may arise in applications such as computer motherboards in which hundreds of amperes (dc) at low voltage are passing through an EMI filter. In this case the series resistance of the transformer will be a problem and the cancellation windings for the capacitors will need to be very large.

2.6 Conclusion

Filters are important for attenuating ripple, eliminating electromagnetic interference (EMI) and susceptibility, improving power quality, and minimizing electromagnetic signature. Capacitors are critical elements in such filters, and filter performance is strongly influenced by the capacitor parasitics. This chapter introduced a new design technique that overcomes the capacitor parasitic inductance that limits filter performance at high frequencies. This new technique is based on the application of coupled magnetic windings to effectively cancel capacitor parasitic inductance, while introducing inductance in filter branches where it is desired. After treating the underlying basis of the technique, this chapter addressed its application to the design of both discrete filters and novel integrated L-section filter components. Numerous experimental results are provided that demonstrate the high performance of the approach by improving attenuation by typically 20 to 30 dB at high frequencies in both discrete filters and integrated filter components across a variety of capacitor sizes and types.

Chapter 3

Filters with Inductance Cancellation using Printed Circuit Board Transformers

3.1 Introduction

Capacitors suffer from both resistive and inductive parasitics. At high frequencies, the equivalent series inductance (ESL) of a capacitor dominates its impedance, limiting its ability to shunt high-frequency ripple current. For example, large electrolytic capacitors often start to appear inductive below 100 kHz, large valued film capacitors become inductive in the 100 kHz - 1 MHz range, and small-valued film capacitors and large ceramic capacitors typically become inductive in the 1 - 10 MHz range. Capacitor parasitic inductance has a significant impact on filter performance, resulting in larger, more expensive filters than would otherwise be needed.

This chapter continues to explore the new filter design technique introduced in the previous chapter to overcome capacitor parasitic inductance. The technique is based on the application of coupled magnetic windings to effectively cancel the parasitic inductance of capacitors, while introducing inductance in filter branches where it is desired. This chapter focuses on the use of air-core printed circuit board (PCB) transformers to realize parasitic inductance cancellation of filter capacitors. Methods to design the transformers will be examined. Aspects relating to manufacturable such as part-to-part variation and sensitivity to outside influences, will be examined. As will be shown, the design approach explored here can provide dramatic improvements in filter performance without impacting the filter size or cost.

3.2 Transformer Design

The most critical element in the application of inductance cancellation is the design of the transformer. In order to define what type of transformer is needed, let's examine typical components that will use inductance cancellation filter capacitors. Typical filter capacitors have ESL values that are in the tens of nanohenries, with a small standard deviation. For example, Fig. 3.1 shows the measured ESL of a large number of 0.22 μF X-type EMI filter capacitors (Beyschlag Centrallab 2222-338-24-224 0.22 μF, 275 V_{ac}). The mean ESL of these capacitors is 16.81 nH with a standard

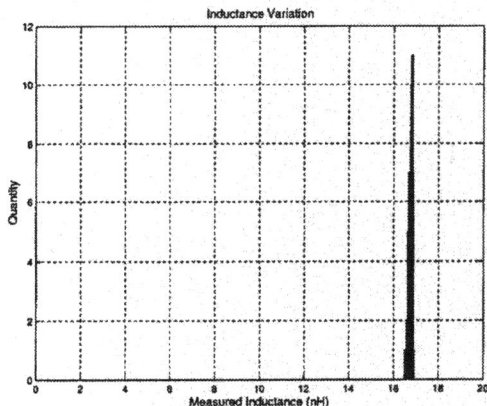

Figure 3.1: A histogram of the parasitic inductances found for Beyschlag Centrallab 2222-338-24-224 X-type capacitors (0.22 μF, 275 V_{ac}). The average value is 16.81 nH, with a standard deviation of 112 pH.

deviation of only 112 pH. Thus, inasmuch as appropriate inductance cancellation magnetics can be realized, tremendous reductions in the effects of capacitor parasitic inductance can be achieved.

Aspects of low inductance transformer design are important for the implementation of inductance cancellation. The inductance and coupling of the magnetic windings must be very precisely controlled in order to accurately cancel the effective inductance of the capacitor. Furthermore, these characteristics must be repeatable from unit to unit, and must be insensitive to operating conditions. Air-core transformers printed directly on the circuit board offer these characteristics. Printed windings provide an extremely high degree of consistency: in the absence of substantial amounts of magnetic material the inductances are purely a function of geometry (making them insensitive to operating condition). Furthermore, to the extent that the PCB space beneath the capacitor can be used to implement the inductance-cancellation transformer, there will be no increase in filter size or cost.

In this section, we address the design of printed PCB transformers for realizing inductance cancellation of filter capacitors. We first consider analytical and computational methods for sizing the printed circuit board windings. We then provide a comparison of winding topologies for inductance-cancelled transformers.

3.2.1 Winding Topology

The two transformer topologies shown again in Fig. 3.2 (which we term end-tapped and center-tapped) are useful for realizing inductance cancelled filters. In order for an end-tapped transformer to be effective the mutual inductance of the two windings must exceed the self-inductance of one of the windings. This requirement usually results in one winding that consists of only one turn and another winding that consists of many turns. Since the first winding has only one turn, the trace width is usually designed to be large in order to minimize shunt-path resistance. As illustrated in Fig. 3.3 , the cancellation term depends on both the self-inductance of the single-turn coil and the mutual inductance of the coils being accurate.

Center-tapped transformers are easier to design. To be effective, a center-tapped transformer only needs to have windings with a controlled mutual inductance. There are no restrictions on

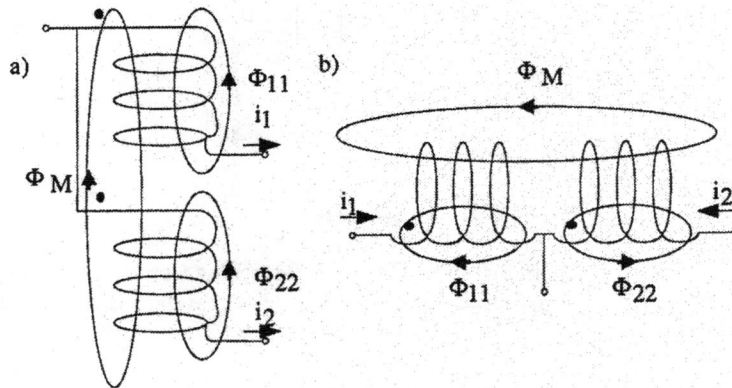

Figure 3.2: (a) An end-tapped and (b) a center-tapped connection of coupled magnetic windings.

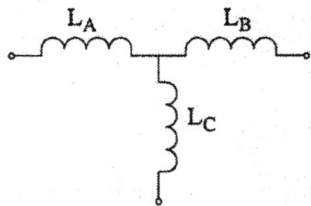

	L_A	L_B	L_C
End-tapped	L_M	$L_{22} - L_M$	$L_{11} - L_M$
Center-tapped	$L_{11} + L_M$	$L_{22} + L_M$	$-L_M$

Figure 3.3: An equivalent circuit model for the configurations of Fig. 3.2.

the relative sizes of the two windings, either in terms of total inductance or impact on shunt-path resistance. Because of this, the center-tapped transformer provides more design flexibility.

The physical size and characteristics of the transformer depend on the choice of winding topology. In an end-tapped transformer the mutual inductance must exceed the self-inductance of the first coil, L_{11}, by the amount of inductance to be cancelled. By contrast, the magnitude of the mutual inductance in the center-tapped topology only needs to equal the cancellation value. The mutual inductance must therefore be significantly larger in an end-tapped transformer design than in a center-tapped design. End-tapped designs thus require more turns (and have higher winding self inductance) than corresponding end-tapped designs. Furthermore the low-inductance winding in an end-tapped design should have a relatively wide trace width so as not to introduce excessive shunt-path resistance. Ultimately, the total board area needed for an end-tapped transformer can be significantly larger than that needed for a center-tapped design.

End- and center-tapped designs also differ in that end-tapped designs tend to result in asymmetric branch impedances (e.g., $L_B > L_A$ in Fig. 3.3), whereas center-tapped designs may be either symmetric or asymmetric. The asymmetry in end-tapped designs arises from the need to have L_{22} much larger than L_{11} such that the mutual inductance will be sufficiently high for reasonable coupling values within the constraint

$$L_{11} < L_M < \sqrt{L_{11}L_{22}} < L_{22} \tag{3.1}$$

The relatively large inductance and asymmetry found in end-tapped designs are not always disadvantageous, particularly if the large branch inductance can be exploited as part of a filter or converter.

To illustrate these effects, both a center-tapped and an end-tapped transformer were designed and compared using the magnetic modeling tool FastHenry [18]. Figure 3.4 shows the layout of these planar transformers. The transformers were designed to compensate for a capacitor parasitic inductance of 21.5 nH, and to provide a dc current path rating of 6 A. The characteristics of the two transformers are listed in Table 3.1. It can be seen that the end-tapped transformer takes up twice the board area of the center-tapped design. Also, for the same negative inductance in branch C (Fig. 3.3), the end-tapped transformer has larger, more asymmetric inductances in the remaining branch paths than does the center-tapped design.

There are also second-order differences between center-tapped and end-tapped winding designs. One design consideration is the sensitivity of the transformer to a nearby metal sheet (e.g., a ground plane or metal chassis). This factor is addressed in a later section, but the results show that center-tapped transformers are somewhat less sensitive to this influence. Another consideration is frequency dependence of the cancellation. As frequency increases the inductance of the coils will change slightly. Specifically, the magnitudes of all the inductances will decrease due to skin and proximity effects. For a center-tapped transformer, the effective negative inductance magnitude will decrease at higher frequencies, while the negative inductance magnitude for an end-tapped transformer can increase at higher frequencies. This occurs when the self-inductance, L_{11}, decreases at a faster rate than the mutual inductance. Thus, winding topology can impact second-order frequency dependencies. If the parasitic inductance of the capacitor is slightly depends on frequency then the high frequency performance of the transformer will be important. Nevertheless, this frequency dependence is very small and its presence has not yet been fully exploited.

The self and mutual inductances of the PCB transformer are, to a small extent, frequency related. The transformers used in Table 3.1 were modeled in FastHenry for a range of frequencies.

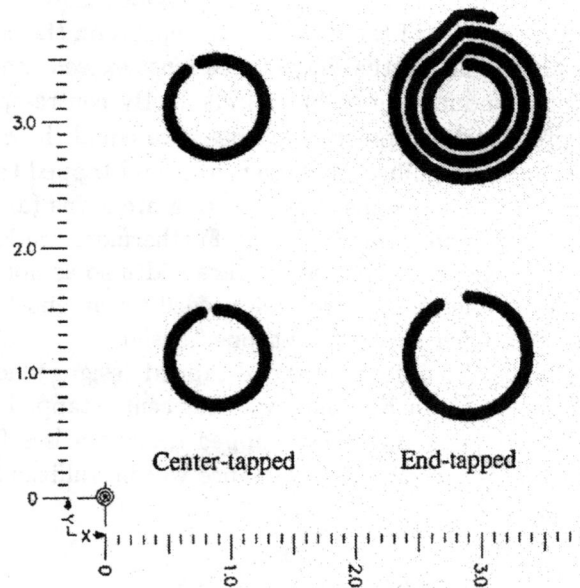

Figure 3.4: Examples of end-tapped and center-tapped transformers. These are the transformers used for Table 3.1. The windings on the left has the top and bottom-side traces of the center-tapped transformer as viewed from the top side. The windings on the right has the top and bottom-side traces of the end-tapped transformer as viewed from the top side. Length is in inches.

	End-tapped	Center-tapped
R_{11}	$1.45m\Omega$	$1.17m\Omega$
L_{11}	46.96 nH	35.71 nH
R_{22}	$4.348m\Omega$	$1.17m\Omega$
L_{22}	251.4 nH	35.74 nH
L_M	68.54 nH	21.44 nH
L_A	68.54 nH	57.15 nH
L_B	182.86 nH	57.15 nH
L_C	-21.58 nH	-21.44 nH
Area	$1.286in^2(829.7mm^2)$	$0.608in^2(392.3mm^2)$

Table 3.1: A comparison of an end-tapped and a center-tapped transformer. R_{11} is the resistance of the first winding (the bottom winding with respect to Fig. 3.4) and R_{22} is the resistance of the second winding (the top winding). L_M is the mutual inductance and L_C is the shunt inductance shown in Fig. 3.3. Area is the maximum are needed by one of the coils. The transformers are rated for a dc-path current of 6 amps.

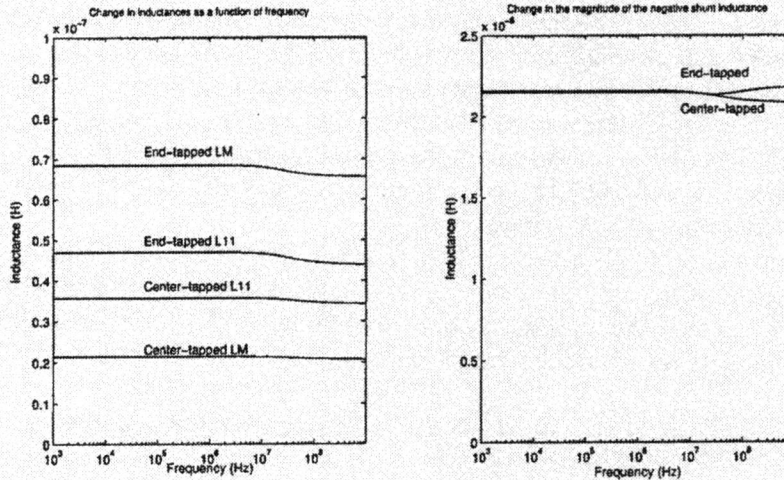

Figure 3.5: Variations in inductances in PCB transformers as a function or frequency. The self and mutual inductances in this example all decrease as a function of frequency. But the shunt path inductance of the end-tapped inductance can be designed to either increase or decrease as a function of frequency.

The self inductances were found to be constant up to 1 MHz and then to drop about 5 % as shown in Fig. 3.5a. This affects end-tapped and center-tapped transformers differently. The shunt path inductance of the center-tapped transformers will decrease in magnitude at high frequency. The shunt path inductance of the end-tapped transformer is related to a self and a mutual inductance, thus depending on the rate at which both these inductances are changing, the shunt path inductance may increase or decrease as a function of frequency. The magnitude of the shunt path inductances are shown in Fig. 3.5b. Therefore the transformers could be designed to change inductance at high frequency by adjusting the thickness and width of the traces. This feature seems to be an exploitable attribute to increase performance at frequencies. However, trace thickness is not readily adjustable and the minute improvements at high frequencies are difficult to measure, thus a thorough investigation into this will not be done at this time.

3.2.2 Inductance Cancellation Winding Design

In order to maximize the benefit of inductance cancellation, the transformer needs to be designed with a high degree of accuracy. Here we outline analytical, computational, and empirical means for designing and refining PCB inductance cancellation transformers. The first step is to determine the inductance to cancel. The parasitic inductance due to the capacitor itself should be measured, and the additional stray inductance associated with the connection of the capacitor in the circuit should be measured or estimated. The transformer is designed to compensate for the total inductance in the capacitor branch.

The transformer can be designed after the inductance to cancel has been determined. Several characteristics of the transformer can be defined based on the following properties: the coils will have a height determined by the standard weight of copper used on the board; the trace widths are sized to handle the AC and DC currents for the application, routinely fifty mils to several hundred mils (~1-5 mm) for traces carrying high currents; the spacing between layers will be equal to the

31

layer or board thickness, normally 62 mils (1.6 mm) for two-layer boards; the additional board area used by the transformer can be minimized by placing it under the capacitor.

In order to design an appropriate transformer, one needs to predict the self and mutual inductance of the printed windings. We have explored three methods for calculating the inductances of a specified geometry. The first method exploits analytical expressions based upon the electromagnetic system in question. The second method uses empirical formulas derived from measured data. The last method employs numerical techniques.

Several papers have been published (e.g., [30,31]) which derive formulas for determining the self and mutual inductances of flat circular loops of various diameters in a system without magnetic material. We have found that the most accurate prediction for this application is that of [31]. The formula for mutual inductance of circular traces is:

$$M_{12} = \frac{\mu_0 \pi}{h^2 ln\left(\frac{r_2}{r_1}\right) ln\left(\frac{a_2}{a_1}\right)} \int_0^\infty S(kr_2, kr_1) S(ka_2, ka_1) Q(kh) e^{-k|z|} dk \tag{3.2}$$

$$S(kx, ky) = \frac{J_0(kx) - J_0(ky)}{k} \tag{3.3}$$

$$Q(kh) = \frac{2}{k}\left(h + \frac{e^{-kh} - 1}{k}\right) \tag{3.4}$$

where h is the copper thickness, r_2 and r_1 are the outer and inner radii of coil 1, a_2 and a_1 are the outer and inner radii of coil 2, and z is the vertical displacement of the loops. J_0 is a Bessel function of the first kind with order 0. M_{12} is the mutual inductance between coils 1 and 2. The self-inductance is the mutual inductance of a coil with itself, i.e., the radii are equal and z is set to zero.

Empirical formulas are also commonly used for determining inductances (e.g., [32–35]). These formulas generally share a common form, but have various constants that differ depending on the characteristics of the coils that were studied in developing the models. We have developed empirical formulas for the self and mutual inductances of planar rectangular (spiral) coils of a size range that is typical for inductance cancellation windings. Rectangular geometry windings are of interest because they are easy to lay out on a printed circuit board. The empirical formulas described in Appendix B are based on numerical predictions, and have been validated with experimental results. They enable fast approximate sizing of rectangular windings.

Another method to obtain inductance for arbitrary winding patterns is to use a three-dimensional field solver such as INCA3D from CEDRAT or a freeware program, FastHenry [18]. FastHenry, the tool used here, can calculate the self and mutual inductances of any three-dimensional air-core winding geometry. These programs allow for arbitrary winding patterns and provide fairly accurate results.

To design an inductance cancellation transformer we use either empirical or analytical formulas to develop a coil design. The design is then refined using numerical computational tools such as FastHenry. The formulas for inductances provide a quick method to obtain reasonably accurate designs and show how various parameters of the windings affect the inductances. The numerical software then provides a greater degree of accuracy for the implementation. A comparison of the results from various methods for an example coil pair is shown in Table 3.2. The difference between the measured and calculated results is more likely a constant measurement error rather that an error in the Hurley and FastHenry results. The fact that these two methods result in nearly identical

32

Source of Inductance Value	Inductance		
	Circular Coil 1	Circular Coil 2	Mutual
Measured	46.25 nH	44.7 nH	23 nH
Zierhofer [30]	36.8 nH	36.8 nH	46.6 nH
Hurley [31]	58.2 nH	58.2 nH	38 nH
FastHenry [18]	57.4 nH	57.4 nH	37 nH

Table 3.2: Comparison of values of inductance calculated and measured using a variety of sources. The transformer consists of two single turn coils with radii (to the trace center) of 510 mils (13 mm) and trace width of 100 mils (2.54 mm).

results, and in similarly good results from other experiments, leads me to believe that numerical software provides the greatest accuracy for this application.

One topic that cannot be ignored in the design of PCB based inductance cancellation circuits is the role of the interconnections on the parasitic shunt inductance. The connections to and from the transformer can add a considerable amount of inductance, at least on the scale of the residual inductance. The final step in the design of the transformer always includes modeling these interconnections with the inductance calculation program. Oftentimes the inductance cancellation design must go through a design refinement iteration.

3.2.3 Design Refinement

Ideally, given good measurements of the parasitic inductance and an appropriate transformer design, the system should have little or no inductance in the shunt path. Unfortunately, the methods used to determine the parasitic inductance and the transformer's T-model parameters are often in error by several nanohenries, a significant amount in such systems. To provide the best cancellation, we routinely add an additional iteration in which a prototype is developed and design refinements are made. We have successfully used four methods to refine transformer design. Here we describe each of these in turn.

The first method of experimental refinement involves fabricating a prototype board incorporating a transformer with multiple tap points. Instead of connecting the cancellation transformer to the remainder of the circuit, the windings are terminated with a set of jumpers or pads at different positions. The connection of the transformer to the remainder of the circuit is made afterwards, with the best connection point determined empirically. The connection method is also important: it should be done in a manner that can be replicated with a printed circuit trace in the final design (e.g., by using a wire or foil link). After the best termination position has been determined, the final board can be built with a printed connection trace.

The second refinement method also involves fabricating a prototype board. In this case, a range of printed transformers are fabricated, each with different predicted characteristics (e.g., with predicted negative inductances spaced evenly over a range.) The transformers are designed with identical interconnects to the external circuit, but with slightly varying coil dimensions (e.g., as determined using (3.2)-(3.4)). One of the fabricated sets will provide the best performance, while the others either over- or under-compensate the parasitic inductance. The best transformer and its interconnect pattern (including ground plane, etc.) is then used in the actual design. The difference between the performance of the various transformer design proceeds in a logical fashion.

Source of Inductance Value	Inductance		
	Circular Coil 1	Circular Coil 2	Mutual
Measured	46.25 nH	44.7 nH	23 nH
Zierhofer [30]	36.8 nH	36.8 nH	46.6 nH
Hurley [31]	58.2 nH	58.2 nH	38 nH
FastHenry [18]	57.4 nH	57.4 nH	37 nH

Table 3.2: Comparison of values of inductance calculated and measured using a variety of sources. The transformer consists of two single turn coils with radii (to the trace center) of 510 mils (13 mm) and trace width of 100 mils (2.54 mm).

results, and in similarly good results from other experiments, leads me to believe that numerical software provides the greatest accuracy for this application.

One topic that cannot be ignored in the design of PCB based inductance cancellation circuits is the role of the interconnections on the parasitic shunt inductance. The connections to and from the transformer can add a considerable amount of inductance, at least on the scale of the residual inductance. The final step in the design of the transformer always includes modeling these interconnections with the inductance calculation program. Oftentimes the inductance cancellation design must go through a design refinement iteration.

3.2.3 Design Refinement

Ideally, given good measurements of the parasitic inductance and an appropriate transformer design, the system should have little or no inductance in the shunt path. Unfortunately, the methods used to determine the parasitic inductance and the transformer's T-model parameters are often in error by several nanohenries, a significant amount in such systems. To provide the best cancellation, we routinely add an additional iteration in which a prototype is developed and design refinements are made. We have successfully used four methods to refine transformer design. Here we describe each of these in turn.

The first method of experimental refinement involves fabricating a prototype board incorporating a transformer with multiple tap points. Instead of connecting the cancellation transformer to the remainder of the circuit, the windings are terminated with a set of jumpers or pads at different positions. The connection of the transformer to the remainder of the circuit is made afterwards, with the best connection point determined empirically. The connection method is also important: it should be done in a manner that can be replicated with a printed circuit trace in the final design (e.g., by using a wire or foil link). After the best termination position has been determined, the final board can be built with a printed connection trace.

The second refinement method also involves fabricating a prototype board. In this case, a range of printed transformers are fabricated, each with different predicted characteristics (e.g., with predicted negative inductances spaced evenly over a range.) The transformers are designed with identical interconnects to the external circuit, but with slightly varying coil dimensions (e.g., as determined using (3.2)-(3.4)). One of the fabricated sets will provide the best performance, while the others either over- or under-compensate the parasitic inductance. The best transformer and its interconnect pattern (including ground plane, etc.) is then used in the actual design. The difference between the performance of the various transformer design proceeds in a logical fashion.

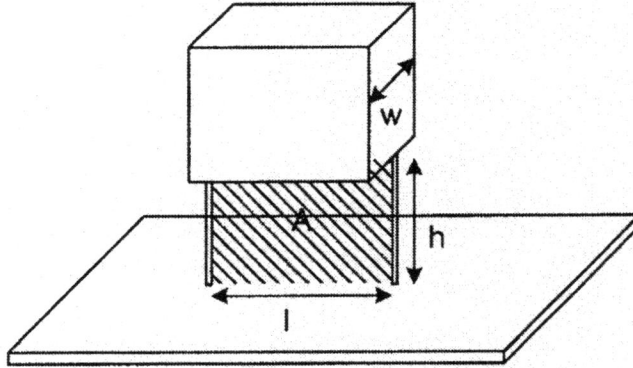

Figure 3.6: The parasitic inductance created by lifting the capacitor off the board can be approximated using l, h, w, and μ_o.

The inductances caused by interconnections and other sources that are hard to model will be constant across the range of transformers. Therefore once several designs are tested, interpolation can be used to pinpoint possible better results.

The third refinement method also uses the same set of test transformers as used above. The only difference between the capacitors and transformers used in the second method is the size of the coils, whose inductances can be calculated. Assume that the amount of parasitic inductance in the capacitors and the interconnects is constant over all the test cases and that the system with the best performance has zero shunt path inductance. Thus, by comparing each test transformer to the test case with the best performance, the amount of shunt path inductances in all the systems is known. When a similar capacitor is used in another circuit with a different type of interconnection or winding pattern there may be too much or too little shunt path inductance. The performance of this circuit can be compared to the set of test transformers to estimate the amount of shunt path inductance in the circuit. This method is simply an easy way to measure the shunt path inductance, after which the transformer can be redesigned to compensate for the error in shunt path inductance.

The final refinement method is useful when a first-pass design overcompensates the shunt inductance, resulting in a net negative shunt-path inductance. If the capacitor is elevated off the board (i.e., with increased lead length) the shunt path inductance will increase. At some height off the board, h, the shunt path inductance will be minimized. The additional inductance introduced by the leads can then be estimated [36]:

$$\Delta L = \mu_0 \frac{hl}{w} \tag{3.5}$$

where the variables w, h, and l are as defined in Fig. 3.6. Once the error has been quantified the transformer can be redesigned for an incremental change in shunt path inductance of the desired amount.

34

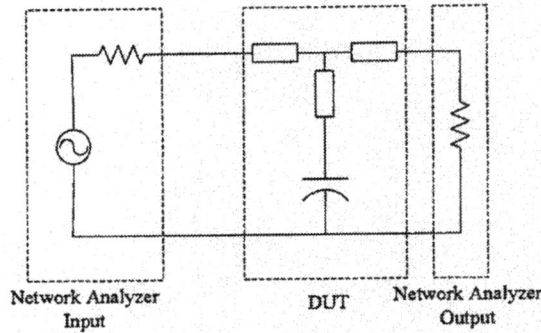

Figure 3.7: An experimental setup for evaluating filters and components incorporating an Agilent 4395A network analyzer. The device under test (DUT) comprises of a filter capacitor and a printed circuit board inductance cancellation transformer with their interconnects.

3.3 Experimental Evaluation and Testing

This section presents an experimental evaluation of the proposed design approach and explores the impact of second-order effects on the repeatability and sensitivity of filters with inductance cancellation. It will be shown that filters incorporating printed PCB cancellation windings can be made highly consistent. Furthermore, the sensitivities to ground planes and other nearby conductors and to magnetic materials are quantified, and found to be low for a wide range of conditions.

Figure 3.7 shows a test setup for evaluating the efficacy of inductance cancellation. The device under test (DUT) may be a capacitor with an inductance cancellation transformer or may be an entire filter. The DUT is driven from the 50 Ω output of a network analyzer. As the driving point impedance of the DUT is always much less than the output impedance of the network analyzer, the drive essentially appears as a current source. The response at the output port of the DUT is measured across the 50 Ω input of the network analyzer. The test thus measures the output response of the DUT due to an input current. This is an effective measure of the attenuation capability of the DUT.

3.3.1 Comparison of Systems with Different Shunt Path Inductances

To validate the proposed approach and to illustrate how variations in inductance cancellation impact performance, a set of center-tapped transformers having a wide range of mutual (cancellation) inductances were designed (Fig. 3.8) for an X-type filter capacitor (Beyschlag Centrallab 2222-338-24-224 0.22μF, 275 V_{ac}). The design approach of Section 3.2 was followed. The transformers were designed using (3.2) and the designs were verified with the program FastHenry. The effects on filter performance of different amounts of mutual (cancellation) inductance is shown in Fig. 3.9 (using the test setup of Fig. 3.7). The highest curve in both Fig. 3.9a and 3.9b represents the capacitor used in a typical fashion without inductance cancellation. The curves in Fig. 3.9a are the results for transformers with mutual inductances that are less than the total shunt-path inductance (between 6 nH and 26 nH at intervals of 4 nH each). Figure 3.9b shows the results of having a mutual inductance that is too large; these curves are the results of transformers with mutual inductances between 26 nH and 32 nH with intervals of 2 nH. Note that the measured parasitic inductance of the capacitor alone (Fig. 3.1) is 10 nH lower than the design value of the cancellation transformer

35

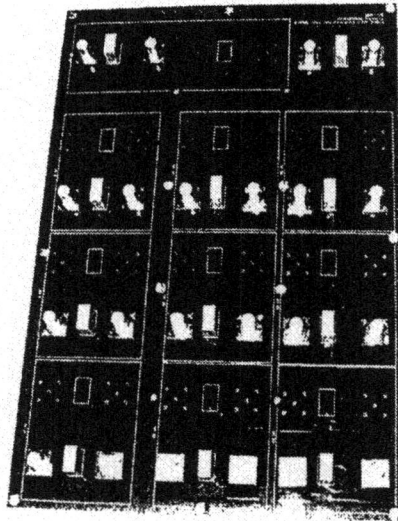

Figure 3.8: A test comparison board. A capcitor without inductance cancellation is in the upper right corner. Every other layout has a transformer with a different shunt path inductance.

that provides the best performance ($\sim$ 26 nH). This reflects additional interconnection inductance along with limitations in our ability to precisely predict inductance.

We may conclude from these results that correct implementation of inductance cancellation can provide large performance improvements (more than a factor of 10 (20 dB) improvement in attenuation across a wide frequency range for the cancellation transformer with the best matching). To achieve this, however, parasitic inductance estimation (measurements and calculations) must be done very accurately, and must include all interconnect inductance in the desired configuration. This figure also shows that 15 dB of improvement in attenuation can be achieved even with a 4 nH error in inductance cancellation.

3.3.2 Test of Part to Part Variation

In order for the proposed inductance cancellation technique to be practical, the cancellation must be highly consistent. It has already been demonstrated that the standard deviation of the equivalent series inductance of off-the-shelf film capacitors is very small (similar results were found for electrolytic capacitors in Chapter 2). Furthermore, one can reasonably expect that printed air-core transformers will provide very repeatable inductances, since inductance is only a function of tightly controlled geometric factors in this case. Here we demonstrate that the performance of inductance cancellation using printed circuit board transformers is very consistent. Six inductance-cancelled filters comprising nominally-identical PCB transformers populated with randomly-selected X capacitors (of the type used above) were constructed. Figure 3.10 shows the performance of these filters in the test setup of Fig. 3.7. The top curve shows the response with a capacitor alone (without use of inductance cancellation). The lower six curves show the performance of the six filters incorporating printed cancellation windings. The performance of the six inductance-cancelled filters are nearly identical, with variations among units of less than 3 dB at frequencies up to 30

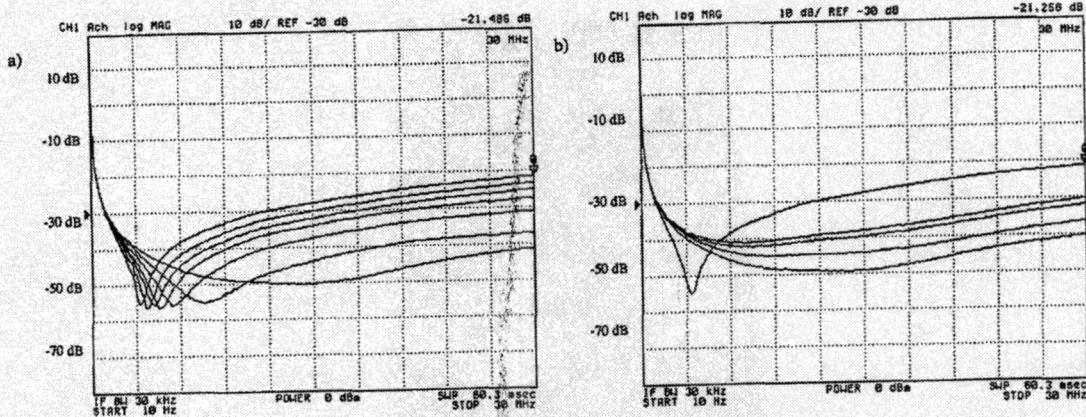

Figure 3.9: The highest curve in both Fig. 3.9a and 3.9b represents the performance of a capacitor without any inductance cancellation in the test setup of Fig. 3.7. The curves in Fig. 3.9 are the results with center-tapped transformers having mutual inductances between 6 nH and 26 nH at intervals of 4 nH each. Figure 3.9b shows the results of having a mutual inductance that is too large; these curves are the results with transformers having mutual inductances between 26 nH and 32 nH at intervals of 2 nH.

MHz. In every case, more than a factor of 10 (20 dB) improvement in attenuation is achieved over a capacitor alone across a wide frequency range. It may be concluded that the proposed approach can achieve large and very consistent improvements in filtering performance.

3.3.3 Interchangeability

Another characteristic that would benefit the practical use of the proposed inductance cancellation technique is if the capacitor can be replaced with similar (but not identical) parts. Typical filter designs will admit alternative capacitors from different sources. The proposed technique will work best when the replacement part exhibits the same parasitic inductance. Figure 3.11 shows the performance of a filter using three different types of X capacitors using the test setup of Fig. 3.7. The three capacitors have the same pin spacing, but different packages, so they are not identical replacement parts. The inductance cancellation transformer was designed for the Beyschlag Centrallab 222-338-24-224 capacitor; the Panasonic capacitors (ECQ-U2A224MG and ECQ-U2A224ML) have about 4 nH more inductance and therefore the performance of the systems with these capacitance differ from that of the original. Nevertheless, despite the fact that these are not identical replacement parts, performance is still greatly improved as compared to the uncancelled case. It may be concluded that the proposed approach is at least reasonably tolerant of component replacement and second sourcing.

3.3.4 Ground Plane Spacing

Ground planes are often used in high-performance power circuits and filters. Clearly, however, a ground (or other) plane should not usually be placed under a printed inductance cancellation transformer. Furthermore, the edge of any plane should be placed some radius away from the coil so that it will not interfere with the coil coupling (and will allow a flux return path). To quantify the size of the keepout region needed around a PCB cancellation transformer to prevent changes in

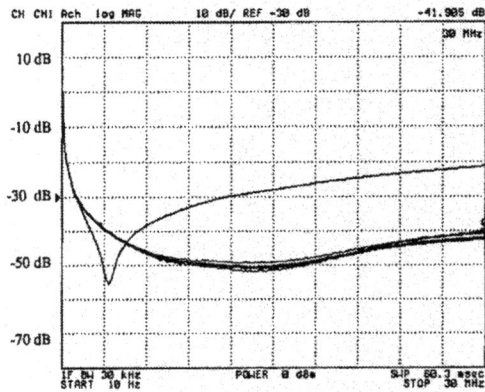

Figure 3.10: Variation test curves. The higher curve is a capacitor alone (no cancellation). The lower curves are results with six nominally identical circuits incorporating printed cancellation windings. Note that the curves are all within 3 dB of each other.

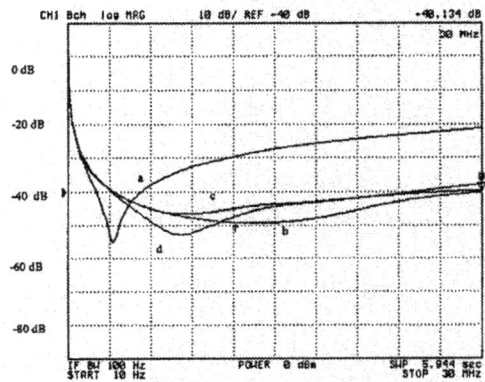

Figure 3.11: Performance of an inductance cancellation filter using different types of capacitors, as measured using the test setup of Fig. 3.7. Trace (a) is the performance of a Beyschlag Centrallab 2222-338-24-224 capacitor without inductance cancellation. Traces (b), (c), and (d) use the Beyschlag Centrallab 2222-338-24-224 capacitor and the Panasonic capacitors ECQ-U2A224MG and ECQ-U2A224ML. The Panasonic capacitors have the same rating (0.22μF , 275 V_{ac}) and pinouts, but different packages.

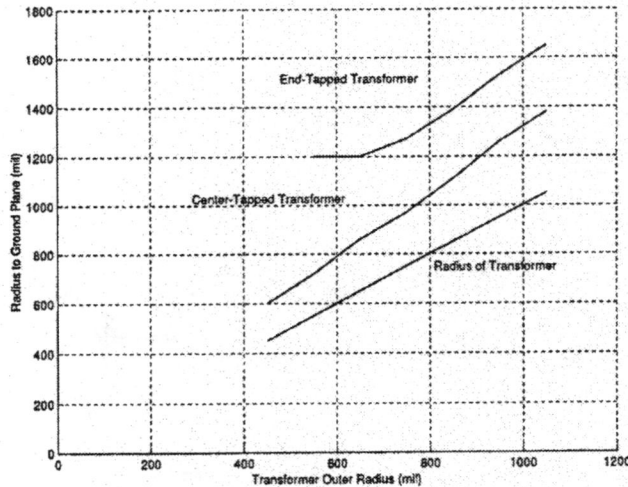

Figure 3.12: The amount of spacing needed between the transformer and the groundplane to ensure that the shunt path inductance is within 1dB of its desired value. The lowest curve represents the outer edge of the transformer coil. The spacing needed for a center-tapped transformer is about 33% larger than the outer radius of the transformer. The radius needed for end-tapped designs is larger.

its performance, we simulated (circular) transformers of various outer diameters and with various ground plane configurations in FastHenry. Each ground plane consisted of a plane with a circular hole of specified radius centered on the cancellation transformer. (Center-tapped transformers were realized as two coils with one turn each, while end-tapped transformers were realized as a three turn coil and a one turn coil. In each case, the coils were on different layers with a 62 mil (1.6 mm) spacing. We then identified the minimum radius of the ground-plane keepout region needed such that the effective negative inductance provided by the printed transformer was within 1 dB of the value achieved without a ground plane present. The results of this study are illustrated in Fig. 3.12. For a circular center-tapped transformer of the dimensions considered, the radius of the ground plane keepout region should be at least 33% larger than the outer radius of the cancellation transformer. The results are somewhat more complicated for end-tapped designs, but generally a larger keepout radius is needed, as per Fig. 3.12. Subsequent experimental measurements confirmed that the proposed keepout regions are sufficient for maintaining the desired level of performance.

To validate the effectiveness of FastHenry for predicting groundplane effects, two closely related systems were fabricated, tested, and simulated. Inductance cancellation transformer and ground planes were fabricated with at least 0.5 mils precision. The transformers were each comprised of 100 mils 2.5 mm wide traces with an outer radius of 368 mils (9.35 mm) with two coils having 2 turns and 1 turn. In the first system, a keepout radius of 568 mils (14.43 mm) was used, while in the second system a keepout radius of 468 mils (11.89 mm) was used. The effective shunt-path inductances of each of these systems were estimated using the test setup of Fig. 3.7. Estimates were made by measuring the voltage response and fitting the data while neglecting the effect of the series-path inductances. At several frequencies above 15 MHz (A frequency in which the shunt path inductances dominates the impedance) the inductance of the two systems were calculated and the difference in inductances were recorded. Table 3.3 lists the results of this experiment.

Freq	Measured Output Voltage for Keepout R= 568 mils	Measured Output Voltage for Keepout R= 468 mils	Corresponding difference in Inductance (nH)	FastHenry Predicted Predicted Difference in Inductance (nH)
15 MHz	-51.80 dB	-50.26 dB	1.181	1.19
20 MHz	-47.60 dB	-46.31 dB	1.187	1.26
25 MHz	-43.943 dB	-43.06 dB	0.682	1.31
30 MHz	-42.938 dB	-41.65 dB	1.349	1.35

Table 3.3: Experimental verification of FastHenry simulations of groundplane. Two layouts are tested with all features identical except that the keepout circle will have radii of 568 and 468 mils (14.4 and 11.9 mm). This experiment compares the difference in inductance for both simulated and measured results.

The measured and simulated absolute inductances vary because the simulation does not take into account any of the interconnection inductances. However, the difference in inductance between the two keepout radii found in simulation and experiment are quite close. One may thus conclude that the FastHenry prediction of ground-plane effects is sufficient for design purposes.

3.3.5 Effect of Nearby Magnetic or Metallic Material

The flux path in an air-core transformer is not as well defined as in a transformer with a ferrite core. The presence of a large sheet of metal in the space directly above or below the transformer may alter the flux path. (An outer circuit enclosure or other large metallic object in close proximity to the transformer could have this effect, for example.) We have used simulation and experimental measurements to study the effects of such external planes on PCB cancellation transformers. End- and center-tapped transformers similar to those shown in Fig. 3.4 were simulated in the presence of a plane of metal at a fixed distance below the board. The results of this study are illustrated in Fig. 3.13. where the magnitude of the negative equivalent inductance is plotted against the distance between the board and the metal plane. Three cases are presented. In one case, (on the highest curve)the plane approaches either coil of the center-tapped transformer. In the second case (on the middle curve) a plane is positioned near the low inductance coil of the end-tapped design, and in the third case (on the lowest curve) the plane is near the high inductance coil of an end-tapped design. This study demonstrates that the center-tapped configuration is less sensitive to the presence of an external plane. Furthermore, it shows that for an end-tapped transformer, the coil with less inductance can be placed closer to an external plane.

A simulation-based test, similar to that of Fig. 3.12, was carried out to show the effects of the presence of an external plane. In this test the distance to a metal plane beneath a transformer is varied so that the effective negative inductance of the transformer decreases by 1 dB. The test considered center-tapped transformers that consists of one turn on each side of the board and with various outer diameters. End-tapped transformers for this test had three turns on one side and a single turn on the other. The outer radii of both coils are equal and the trace width is 100 mils. As seen in Fig. 3.14, the center-tapped designs are less sensitive to the presence of an external plane. Furthermore, for end-tapped transformers the coil with less inductance can be placed closer to an external plane.

It is reasonable to expect that the presence of magnetic material near the air-core transformer

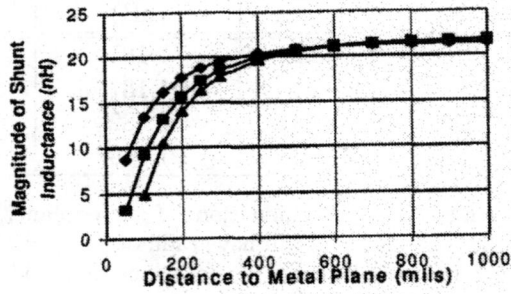

Figure 3.13: The effect of a metal plane on the effective negative shunt inductance. The results are for the transformer designs used in Table 3.1.

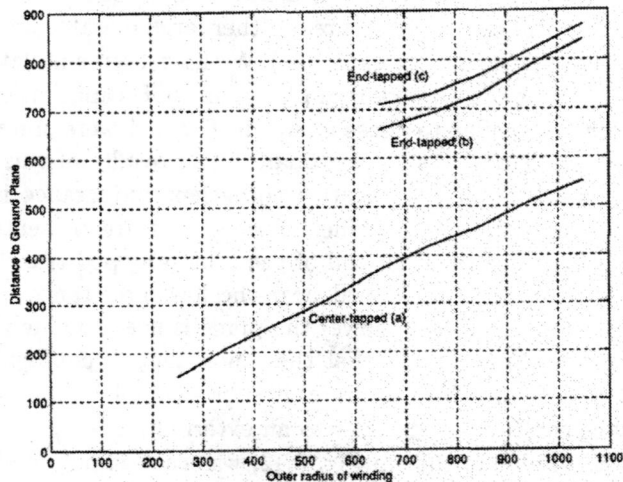

Figure 3.14: Simulation based calculations of the distance between a metal plate and the transformer needed to perturb the shunt path inductance by 1 dB. The curves show the performance of a center-tapped transformer (a) and end-tapped transformers with a plate near the low inductance turn (b) and the high inductance turn (c).

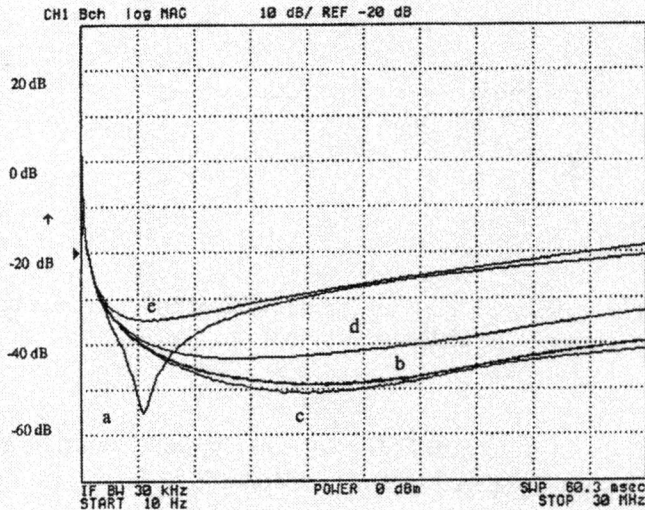

Figure 3.15: The effects of magnetic material on the performance of inductance cancellation. Signal (a) shows the performance of a capacitor with no inductance cancellation windings, while signal (b) shows the performance with cancellation. When a ferrite core is placed adjacent to the capacitor and air-core windings there is not much change (c). The maximum amount of interference occurs when the core impinges directly on the windings. When the core is approximately 100 mils from the board, directly over the coils (d) the performance drops. When the core is placed directly over the windings, inductance cancellation is completely ineffective (e).

could also adversely affect its performance. The presence of material with permeability other than μ_o near the core will influence the flux patterns and may change the inductances in the transformer. Fig. 3.15 shows the effect of magnetic material (type 3F3) when it is placed next to a device under test in the test setup in Fig. 3.7. In this case the transformer used is a 1 turn by 1 turn center-tapped transformer with radius 325 mils (8.3 mm) and a trace width of 100 mils (2.54 mm) on a 52 mil thick pc board. The figure shows that the performance of the transformer is only affected when the magnetic material is placed over the windings and that the amount of influence is related to the distance the magnetic material is from the board. The conclusion of this empirical test is that having magnetic material on the same board and close to the capacitor and windings is not an issue, as long as the material does not impinge directly on the transformer. Also, if the system is placed near other boards (e.g. in a rack) some spacing (in this case 200 mils (5.08 mm)) is needed if magnetic material will be positioned directly below the air-core transformer. Based on these results, we do not anticipate that this issue is a significant problem.

3.4 Design and Evaluation of an EMI Filter

A capacitor with an inductance cancellation transformer is a two-port filter rather than a (one-port) capacitor. The use of an inductance cancellation transformer will benefit some applications and not others. EMI filtering is one application where this technique excels. In addition to improving capacitor performance, the branch inductances introduced by the transformer serve to enhance filtering performance by increasing series-path impedance. This section explores the design of an

42

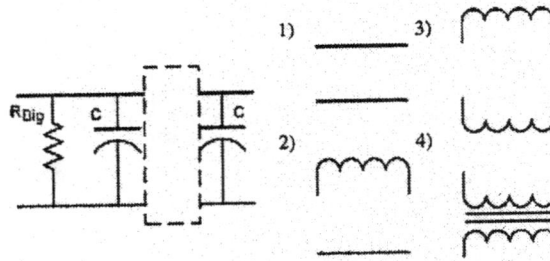

Figure 3.16: The EMI filter under test. The box can be replaced by any of the four connections shown. The filter can also use capacitors with or without inductance cancellation.

inductance cancellation transformer for an EMI filter application. The performance of a filter using inductance cancellation is then compared to a conventional implementation.

Figure 3.16 shows a structure that can be used to realize a variety of filters. With a direct connection between the two stages, the two capacitors appear in parallel. If an inductor is used in one branch, a pi filter is formed, and if inductors are used in both branches a split-pi filter results. If a common-mode choke is used (as is done in many ac applications), one gets common-mode filtering from the choke, and differential mode filtering from the capacitors and the (relatively small) choke leakage inductance. Here we consider the effect of using inductance cancellation on the (differential-mode) capacitors of Fig. 3.16. To simplify evaluation, we consider purely differential-mode connections of the circuit in Fig. 3.16 (i.e., connections 1 and 2). Nevertheless, the basic results apply to the differential-mode behavior of filters providing both common- and differential-mode filtering.

The procedure in Section 3.2.2 was used to design the transformer for both the capacitors. The inductance formulas (3.2)-(3.4) were used to determine the inductance for center-tapped transformers with a one turn and a two turn coil. The entire circuit was modeled in FastHenry and refined using the procedures in Section 3.2.3. In particular the tap points connecting the transformer to the input can, and were shortened to adjust the inductance. The final design uses a transformer with a trace width of 100 mils (2.54 mm) and an outer radius of 345 mils (8.76 mm). According to the inductance formula (3.2) this transformer has a mutual inductance with a magnitude of 19 nH. The layout for the board is shown in Fig. 3.17, along with the layout of the conventional filter.

The performance of the filter circuit for several filter connections (with and without inductance cancellation) is illustrated in Fig. 3.18. These results were obtained using the test setup of Fig. 3.7, as described previously. Trace (d) of Fig. 3.18 shows the measurement noise floor (the response with the network analyzer input and output both disconnected from the filter). Trace (a) shows the performance of the filter with normal capacitors (i.e., without inductance cancellation) connected in configuration 1 (as shown in Fig. 3.16, i.e., two normal capacitors are connected in parallel). As expected, substantial attenuation is achieved, but it becomes worse above the 3 MHz self-resonant frequency of the capacitors. Trace (b) shows the performance of the same configuration using the design with inductance cancellation. Attenuation is greatly improved (over the case without cancellation) for frequencies above the self resonant frequency of the capacitors, reflecting the benefit of nulling their parasitic inductance. At high frequencies, as much as 40 dB of improvement in attenuation is achieved over the conventional implementation.

Despite the large performance improvement that is achieved, the performance still isn't as good

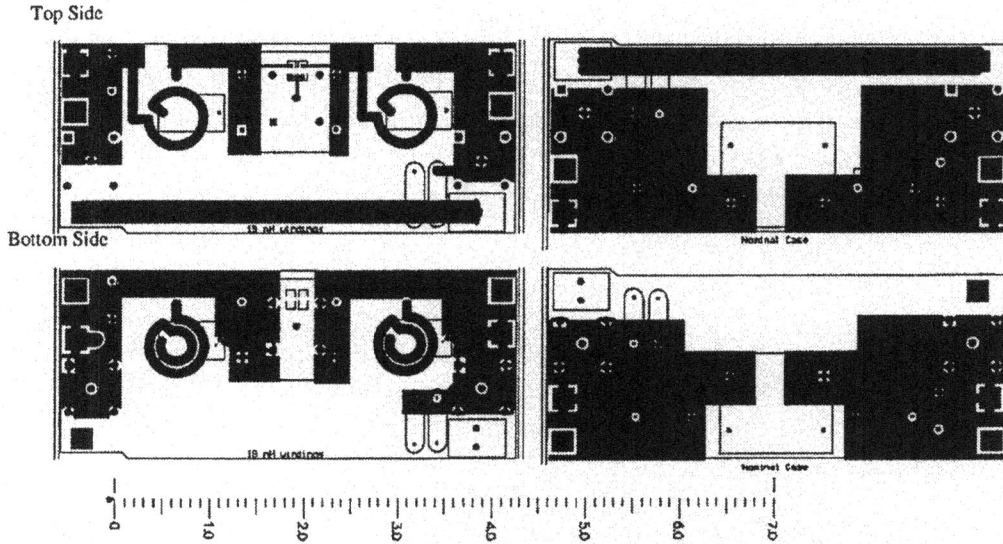

Figure 3.17: The layout of the filter containing capacitors with and without inductance cancellation. The traces in the middle area of the inductance cancellation board are for testing the capacitors and are removed afterwards. The figures on the left show the top and bottom-sides of the filter board with inductance cancellation. The figures on the right show the top and bottom-sides of the filter board without inductance cancellation. Length is in inches.

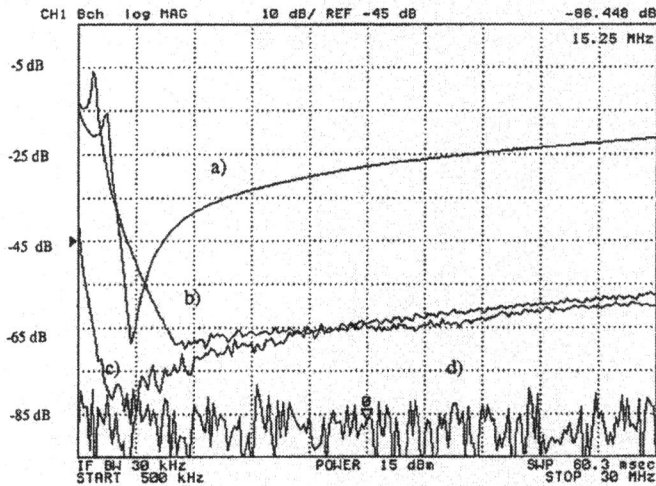

Figure 3.18: The results of the EMI filter test. Signal a is the performance of two normal capacitors. Signal b is the performance of two capacitors with inductance cancellation. Signal c is the performance of two normal capacitors with a single inductor in the series path. Signal d is the noise floor.

as one might anticipate. The rate of improvement of the inductance-cancelled design over the conventional design drops rapidly at about 6 MHz. This occurs because at frequencies above 6 MHz the output response of the filter is dominated by parasitics that entirely bypass the capacitors and cancellation transformers. In fact, for those frequencies, the performance remains unchanged even if the connections between the first and second capacitor networks are removed entirely! Measurements reveal a 1.3 pF parasitic capacitance from the filter input to the filter output that at least partially accounts for this parasitic coupling. Hence, the introduction of inductance cancellation has improved the filter performance to such an extent that small parasitic paths (e.g., associated with layout) are the dominant factor in performance.

Trace c of Fig. 3.18 shows the performance of a pi filter connection (Fig. 3.16 connection 2) without inductance cancellation. As can be seen, filter performance is greatly improved, with the output response falling quickly to the "coupling floor" (the level at which parasitic coupling past the filter components dominates). A 20 μH inductor was selected for the filter. This is the smallest inductor sufficient to drive the output response down to the coupling floor out to 30 MHz. (It was found that a larger inductance did not further increase attenuation at high frequencies, and a smaller inductance provided less attenuation.) Thus, we find that at frequencies above 6 MHz the capacitors with inductance cancellation provide the same attenuation performance as the full pi filter (without inductance cancellation), though the pi filter provides better performance at low frequencies. Depending on the EMI specification and system parasitics, inductance cancellation methods can be as effective as higher-order filtering in achieving high attenuation. With either approach, eliminating filter parasitic coupling (by layout, shielding, etc.) is critical for achieving maximum filter performance.

3.5   Conclusions

Capacitor parasitic inductance often limits the high-frequency performance of filters for power applications. However, these limitations can be overcome through the use of specially-coupled magnetic windings that effectively nullify the capacitor parasitic inductance. This chapter explores the use of printed circuit board (PCB) transformers to realize parasitic inductance cancellation of filter capacitors. Design of such inductance cancellation transformers is explored, and applicable design rules are established and experimentally validated. The high performance of the proposed inductance cancellation technology is demonstrated in an EMI filter application.

This chapter examines and compares the end-tapped and center-tapped transformer for use in pcb transformers and finds that center-tapped transformers were more volumetrically efficient and less sensitive to outside disturbances. Three methods are examined for calculating the inductances for pcb air core magnetics and several methods to experimental refine the design are presented. The calculation methods alone will not provide the degree of precision needed to optimize this technique and to provide the 20 to 30 dB improvement in attenuation that is possible. This chapter examines the effect of inductance cancellation on the pcb board and the entire system. The pcb transformer can be placed under the capacitor and needs a prescribed keep out region for the ground plane. The filter element has consistent performance and the transformer needs only to have some known displacement from other magnetic elements and metal sheets. Last, but not least, the inductance cancellation transformer is "free" in the sense that it provides large performance improvement at little or no increase in size of cost.

Chapter 4

Design of Integrated LC Filter Elements with Inductance Cancellation

4.1 Introduction

As shown in the previous chapter, a filter with inductance cancellation can be implemented with a conventional capacitor and an air-core inductance cancellation transformer implemented in a printed circuit board. This is a highly effective approach, but necessitates significant design effort on the part of the circuit designer. Another approach is the development of an integrated filter element, in which an inductance cancellation transformer is packaged together with a capacitor to form a three (or more) terminal device that provides greatly enhanced attenuation. It is more cost effective for the capacitor manufacturer to do the transformer design for countless applications. The board designer would only have to choose the appropriate product and insert the integrated filter element into the board. The transformer, when integrated into the capacitor, can use volume more efficiently than the pcb transformer and thus transformers with larger inductance and smaller resistance can be formed. Furthermore, the integrated filter can more easily be designed to be insensitive to the presence of ground planes and other magnetics in the circuit. This chapter investigates two types of construction methods for integrated filters elements with inductance cancellation.

4.2 Overview

Here we investigate two methods for constructing integrated filter elements. In the first method a transformer is added around the capacitor. The transformer windings can either be formed as an extension of an electrode winding of the capacitor or from a separate conductor wound about the capacitor. Fig. 4.1 shows the structure of this type of integrated filter element.

The second method of forming an integrated filter element is to add a separately formed transformer to a space near the capacitor. For example a quasi-planar transformer could be added axially to a a capacitor winding. In this case the transformer can be designed using similar methods as a pcb transformer. The capacitor is typically constructed with two terminals. An air core transformer winding made up of a conductor is insulated in a non-conducting material. The transformer

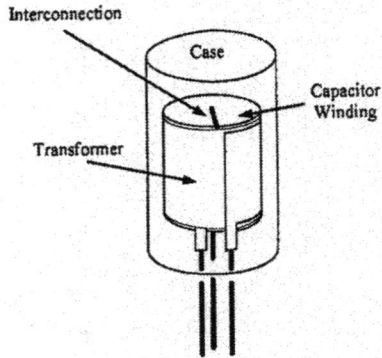

Figure 4.1: An integrated filter element with an air core transformer that is wound around and connected to the capacitor and enclosed in the capacitor casing.

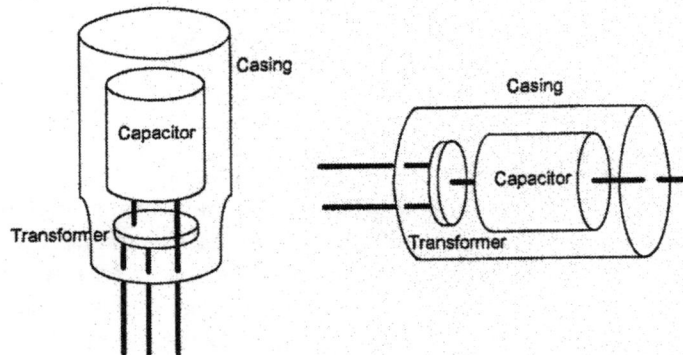

Figure 4.2: Structure of an integrated filter element with a separately constructed transformer connected to the capacitor and enclosed in a case.

can then be connected appropriately to one electrode of the capacitor. The filter element casing will be constructed over the capacitor and the transformer. This type of integrated filter element is illustrated in Fig. 4.2.

4.3 Integrated Filter Element with the Transformer Wound about the Capacitor

Film-foil, metalized film, and several other capacitors are made by taking strips of conductors and insulators and rolling the structure into a tightly wound cylinder. The concept of forming a filter element with a transformer around the capacitor originates from this winding process. With an additional winding procedure - perhaps using the same equipment - the transformer can be formed and connected to the capacitor.

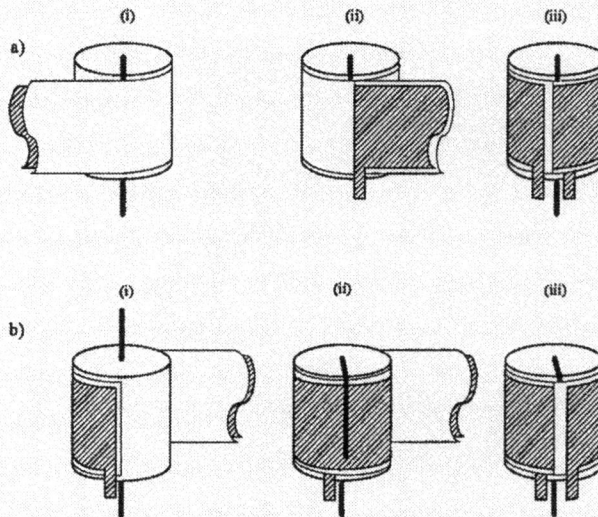

Figure 4.3: Winding methods and connections for (a) an end-tapped inductance cancellation transformer and (b) a center-tapped inductance cancellation transformer.

4.3.1 Design Process

The two different kinds of transformer connections, end-tapped and center-tapped can be used in this application. The type of transformer plays an important role in how the transformer will be wound. Two different winding methods - each suitable for one of the transformer connections - are illustrated in Fig. 4.3. For the end-tapped transformer of Fig. 4.3(a), the transformer winding begins with a connection to one capacitor electrode. The winding can be an extension of the electrode or formed with a separate conductor that connects to the electrode. The transformer is then wound to form N_1 turns, Fig. 4.3ai. At the end of the first winding the transformer will be connected to a terminal of the filter element. The second winding, which may be an extension of the first winding, is wound in the opposite direction (Fig. 4.3aii) for N_2 turns (where N_2 must be greater than N_1), and connected to a second terminal (Fig. 4.3aiii). The third terminal is connected to the other electrode of the capacitor. The first and second terminals refer respectively to terminals A and B of the transformer as shown in Fig. 4.4 with the third terminal as a common. The filter element is bi-directional (i.e., the first and third terminal can be the input port and the second and third can be the output port, or vice versa). Note that the series inductances, L_A and L_B, of Fig. 4.4 are not equal and the relative values of these terms may determine the preferred orientation.

If a center-tapped transformer is used, it cannot be easily formed from an extension of the electrode, since the connection of the capacitor to the transformer is formed at a mid-point in the winding. A separate conductor is therefore wound on the capacitor. The beginning of the transformer will be connected to one terminal of the device. The transformer will be wound N_1 turns around the capacitor (Fig. 4.3bi). At the end of the first winding the transformer will be connected to the electrode of the capacitor (Fig. 4.3bii) and the transformer will be wound an additional N_2 turns in the same direction as the first set of turns. The end of the secondary winding is connected to a second terminal of the filter element (Fig. 4.3biii). The values of N_1 and N_2 can be set in order to make L_A and L_B either equal or at some ratio. The other electrode of

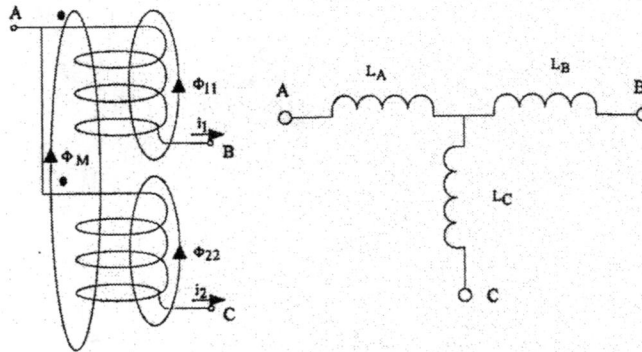

Figure 4.4: The winding description of an end-tapped transformer and the associated T model.

the capacitor is connected to the third terminal of the device.

4.3.2 Fabrication Issues

This construction process for the filter element does have some drawbacks. If the transformer is constructed as a separate winding it can be easily sized for appropriate dc and ac currents. If, however, the capacitor electrode metal is to be used as part of the transformer, one is limited to the conductor thickness used in the body of the capacitor. Usually the metal (electrode) layers inside a capacitor are very thin so that the number of layers and thus surface area can be maximized. The resistance per square of this material can be exceedingly large and therefore the transformer will have excessive parasitic resistance (and low dc current carrying capability). Metallized film capacitors, in which a layer of aluminum is on the order of 50 nm thick (a relatively large amount for this kind of capacitor) will have a resistance of 4.11 Ω for six turns on a capacitor with a height of 2.18 in and a diameter of 0.9 in (e.g., Cornell Dubilier 9351W20K). If the electrode was capable of carrying the current needed, then the end-tapped transformer design, using this electrode to form the transformer winding, would be easier to manufacture than the center-tapped transformer.

A second drawback to this winding method is that two new connections points must be made to the transformer. The transformer will be a thin foil or metallized film and needs tap points to connect to the terminals of the filter element. Reliably making a permanent connection to the transformer, depending on the thickness of the foil or metallization, may be a significant design challenge.

Creapage and clearance requirements between the transformer windings, capacitor windings and the case are also a concern. For example, in an electrolytic capacitor the metal case will be tied to the lowest potential. If the transformer is connected to the high-potential anode (the most desired connection), any contact between the transformer and the case will short circuit the capacitor. This would not be a concern if the cathode terminal was used and the filter element was surrounded by a potential that is approximately ground, but that solution is often unacceptable from the point of view of the circuit designer. Hence, despite its attractive characteristics, insulation and packaging considerations make this approach challenging. The approach described in Section 4.4 seems to be a much more easier winding method with a correspondingly lower manufacturing cost.

49

Number of Turns	Inductance
1	21.7 nH
2	29.5 nH
3	39.7 nH
4	47.5 nH

Table 4.1: Inductances of Metal foil wrapped around United Chemi-Con U767D, 2200 μF, 35 V capacitor for various number of turns.

4.3.3 Design Methods

The self and mutual inductances of the inductance cancellation transformer must be accurately set to make an integrated filter element. This transformer will have a height similar to the height of the capacitor and a very small thickness; this is the opposite of the pcb transformer, which has a small height and a thickness set by the trace width. The inductance cancellation transformer will be wound in close proximity to the metal electrodes and a case which may be a conductor. The presence of nearby conductors will change the inductance of the transformer, thus predictions of the transformer inductances without considering the presence of the capacitor will be overestimations. The internal composition, the thickness of the capacitor insulation and other physical parameter will make analytical formulas for the self and mutual inductances complex. A much simpler method is to determine empirical formulas based on measurements on the capacitor that will be used in the integrated filter element.

The empirical formula is determined by winding an inductor around the capacitor using the backed foil that will be used in the integrated filter element. The inductance between two turns of a foil inductor can be used to estimate the number of turns needed for the inductance cancellation transformer. Several inductors should be formed in this manor and measured. The measured inductances are defined as

$$L_N = N^2 L_{\text{Turn}} + L_{\text{extra}} \tag{4.1}$$

Where L_{extra} is the inductance due to any connections to the impedance analyzer, N is the number of turns, and L_{Turn} is the inductance between two turns. Using this equation for several different values of N will over-define the unknown constants L_{self} and L_{Turn}. These constants can be estimated by finding the least squared error solution.

As an example we consider insulated foil turns (1 mil thick and 500 mils wide) around a United Chemi-Con U767D, 2200 μF, 35 V capacitor. Table 4.1 shows the inductances from measured results. This data can be written in the form of (4.1) as

$$NL = L_N = \begin{bmatrix} 1 & 1 \\ 4 & 1 \\ 9 & 1 \\ 16 & 1 \end{bmatrix} \begin{bmatrix} L_{Turn} \\ L_{extra} \end{bmatrix} = \begin{bmatrix} 21.7 \\ 29.5 \\ 39.7 \\ 47.5 \end{bmatrix} \tag{4.2}$$

Where N is the matrix representing the number of turns corresponding to the test results, L is a matrix of the unknowns, L_{self} and L_{Turn}, and L_N is the matrix of test results. The least squares error solution is

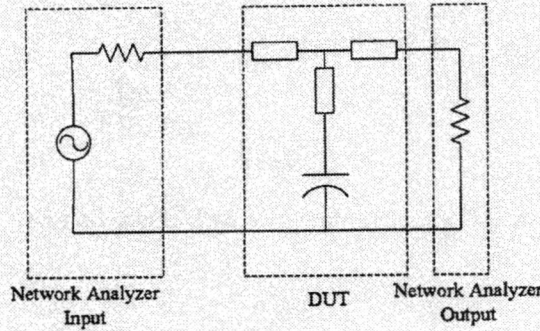

Figure 4.5: The test circuit used for measuring systems with inductance cancellation.

$$L = (N'N)^{-1} N' L_N = \begin{bmatrix} 354 & 30 \\ 30 & 4 \end{bmatrix}^{-1} \begin{bmatrix} 1 & 4 & 9 & 16 \\ 1 & 1 & 1 & 1 \end{bmatrix} \begin{bmatrix} 21.7nH \\ 29nH \\ 39.7nH \\ 47.5nH \end{bmatrix} =$$

$$\begin{bmatrix} L_{Turn} \\ L_{extra} \end{bmatrix} = \begin{bmatrix} 1.6977nH \\ 21.8674nH \end{bmatrix} \tag{4.3}$$

Which gives an estimate of 1.6977 nH as the inductance between two full turns of foil on the capacitor.

The estimation of the self-inductance and the mutual-inductance of the foil windings can be used to develop a transformer for purposes of inductance cancellation. As described in Chapter 2 the amount of parasitic inductance associated with the capacitor needs to be identified. Either a center-tapped or end-tapped transformer can be used, this transformer should have the appropriate negative inductance in one branch of the equivalent T-model. The self and mutual inductance terms used to described the transformer can be approximated as

$$L_1 = N_1^2 L_{\text{Turn}} \tag{4.4}$$

$$L_2 = N_2^2 L_{\text{Turn}} \tag{4.5}$$

$$L_M = N_1 N_2 L_{\text{Turn}} \tag{4.6}$$

After an appropriate transformer is designed the system should be tested to find the exact point on the foil transformer which leads to a system with a minimized amount of shunt path inductance. Assume that the transformer is a center-tapped transformer and that the interior winding is connected as the "input" of the filter element, then the exterior winding is the "output" of the filter element. To test the device a gauge 22 wire is placed to various points on the exterior winding and connected to the output of the filter element until the shunt path impedance is minimized. The test circuit is shown in previous chapters and again in Fig. 4.5.

51

Figure 4.6: A prototype integrated filter element based on a United Chemi-Con U767D, 2200 μF, 35 V electrolytic capacitor and center-tapped cancellation winding. The board used for comparison, consisting of only a normal capacitor, is also shown.

4.3.4 Experimental Validation

Experimental validation of integrated filter elements constructed using turns of copper foil wound around a film and an electrolytic capacitors were presented in Chapter 2. In these examples copper tape insulated with mylar was wound around the exterior of the case of the capacitor. In these tests the windings are placed in parallel and in close proximity to the metallic casing. The film capacitor has a layer of aluminum metalized onto film within millimeters of the surface. The electrolytic capacitor has a metallic case with a thin layer of plastic over it. These tests show that it is possible to create a filter element with inductance cancellation with a small number of turns around a capacitor winding.

This example of an integrated filter element is taken directly from Chapter 2. An integrated filter element based on a 2200 μF electrolytic capacitor (United Chemi-Con U767D, 2200 μF, 35 V) was evaluated. Such capacitors are widely used in filters for automotive applications. The center-tapped cancellation winding was wound 0.25" from the top of the capacitor body with 0.5" wide 1 mil thick copper tape, using 1 mil thick mylar tape for insulation. As stated in the previous subsection the inductance between two terms is estimated to be 1.6977 nH. The measured inductance of the capacitor is 17.7 nH (taken from the histogram in Fig. 2.7). A center-tapped transformer is used in this example therefore the shunt path inductance is $-L_M$. This inductance is given by (4.6) and if the two turn ratios are chosen to be equal then the expected number of turns is 3.2. Experimentally, the tap points are found from these estimates. A 3.25 turn winding was placed on the capacitor body and tapped, followed by a continued 3.125 turn winding. The test board setup is shown in Fig. 4.6

Test results using the experimental setup of Fig. 4.5 are shown in Fig. 4.7. The impedance of the capacitor alone begins to rise in the vicinity of 100 kHz due to ESL effects. By contrast, the

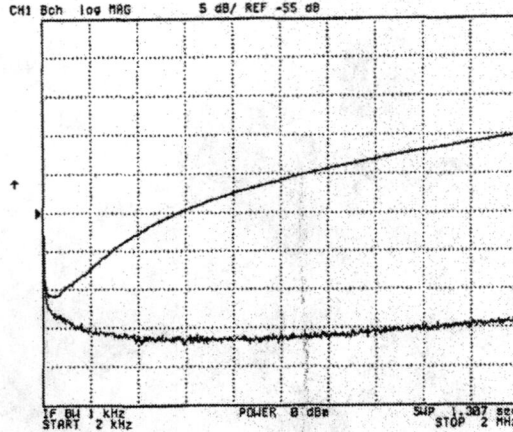

Figure 4.7: Comparison of a United Chemi-con U767D, 2200 μF electrolytic capacitor to the corresponding prototype integrated filter element.

integrated filter element continues to attenuate the input out to much higher frequencies, resulting in more than a 10 dB improvement at 200 kHz, and increasing to more than 20 dB above 1 MHz.

4.4 Integrated Filter Element with Separately-constructed Transformer

The second method to create integrated filter elements does not have the problems associated with winding the transformer around the capacitor. This method uses a separate transformer made from a conductive sheet. This sheet can easily be made of any thickness and is placed in a location that can be easily isolated from the capacitor electrodes. In this construction method the capacitor is constructed normally and the transformer can be easily insulated from the capacitor via spacers or potting material. An inexpensive method of fabricating the transformer is also proposed.

The process steps associated with this construction method are easily automated and the additional cost in making a filter element will be low. Appendix C lists the expected costs of processes and materials needed to construct integrated filter elements using this approach.

4.4.1 Design Process

The first step in the design is to accurately measure the amount of parasitic inductance that needs to be cancelled. The next step is to design an air core transformer with precise inductance values that can be placed inside the casing of the capacitor. The transformer is limited in size by the cross sectional area of the capacitor as shown in Fig. 4.2, but the transformer can have any number of layers. The transformer also needs to be rated for the amount of dc and ac currents it must handle. The thickness of the transformer winding can be adjusted in order to control the winding resistance of the transformer. Therefore, to make filter elements with the same capacitance and different dc-path current ratings, a single type of capacitor can be used with different transformer design. The transformers will have the similar dimensions as transformers made from pcb traces,

Figure 4.8: A patterned 10 mil thick copper sheet suitable for folding into a transformer. The pattern was cut from a sheet using a waterjet cutter.

and can be designed using the methods described in Chapter 3.

Once the design of the transformer and the transformer thickness have been set the transformer can be constructed from a sheet of metal with the appropriate thickness. We consider quasi-planar transformer structures that can be inexpensively fabricated as planar patterns that are folded to form the cancellation transformer. The winding can be made from any conductor and can be formed by a variety of processes. The winding pattern can be stamped from a sheet of metal. The winding can be formed from a wire or thin strip of conductor that is precisely bent, folded, or shaped into the pattern. The winding can be cut directly from a sheet of conductor, for example by a laser cutter or water-jet tool. Fig. 4.8 shows a a planar winding pattern cut from a copper sheet with a waterjet.

The winding is insulated to ensure that it will not make an unwanted electrical connection to itself when formed into a transformer. This insulation can be applied in several ways. The conductor can be sprayed with, painted, printed, or dipped into a chemical that forms an insulation layer. Some metals can be treated to form an insulating layer such as an oxide. An insulating sheet can be placed onto the windings with or without adhesive. These insulation processes can either be carried out on the winding pattern or on the conductor before the pattern is formed (e.g. a sheet of conductor that is coated with an insulation layer can be stamped or an insulation layer can be formed on the winding pattern).

Another task in the design is to reshape the winding into the transformer. For example, a multi-layer, air-core transformer can be made by folding the planar winding pattern into multiple layers. An automated process of taking the winding and folding it into a transformer can be used, for example. Figure 4.9 shows the winding from Fig. 4.8 after an insulation layer of clear foil is added and the winding is folded into a two-layer transformer.

The integrated filter element is finished by connecting the transformer to the capacitor and attaching all the leads. One electrode of the capacitor is connected to an external lead and the other is connected to the transformer. The transformer is connected to two external leads, forming the three terminal filter component. The positions of the connections to the transformer will depend on whether the transformer is a center-tapped or an end-tapped design and on the design of the winding. Connection can be made is a similar manner as those used for capacitor terminals (e.g., welding). To reduce the number of interconnections, the two external terminal leads to the transformer may be a part of the original winding pattern that are formed (e.g., bent) appropriately.

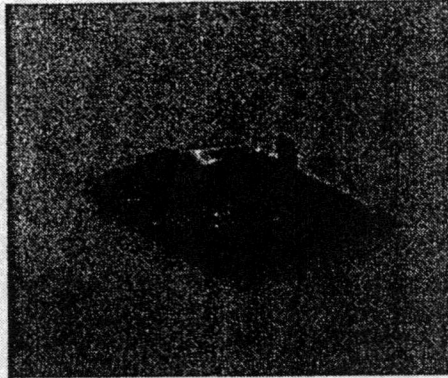

Figure 4.9: The transformer formed by adding an insulation layer of clear foil to the planar winding pattern and folding it.

Both the spacing between the capacitor plate roll (or interconnect) and the transformer, and the spacing between the transformer and the device terminals (e.g. as seen in Fig. 4.2) are important design factors. These spacings are controlled to achieve the desired filtering performance, to make the transformer behavior insensitive to variations in external conditions, and to provide spacing to meet voltage breakdown, creapage, and clearance requirements that may exist. The spacing may be controlled by a number of means. These include 1) placing one or more insulating mechanical spacers (e.g. between capacitor and transformer, transformer and external connection), 2) setting the desired spacing by one or more potting steps (e.g. potting capacitor windings, then positioning and potting the transformer windings), 3) interconnection through one or more fluid proof stoppers (e.g. for an electrolytic filter element), 4) by control of the insulator layer thickness(es) in the transformer winding, or by a combination of these methods.

Figure 4.10 shows the capacitor connected to the transformer. Testing of the system can be made and the system can be evaluated. Adding a transformer with some mandatory separation distance will add extra height (or length for an axial capacitor) to the filter over that needed for the capacitor alone. The extra height corresponds to an increase in the parasitic inductance which is being cancelled. Thus, the transformer needs to be designed to cancel out the parasitic inductance of the capacitor and interconnect to the transformer and the added inductance due to interconnects from the transformer to the outside connection point of the device.

The transformer in this case is encapsulated in a mixture of clear epoxy and is shown in Fig. 4.11 with additional foil leads added.

4.4.2 Experimental Validation

The design steps of the previous subsection were followed to construct several prototype filter elements. The parasitic inductances of a dozen capacitors, 3.3μH X-Type capacitor from BC components (2222 338 24335), were measured, the average value is 22 nH. The transformers for the prototype were designed using FastHenry. The copper thickness is set to 10 mils, and the transformer form factor is limited to that of the capacitor (812 by 1200 mils). Several center-tapped transformers were designed with a range of mutual inductances centered around 22 nH.

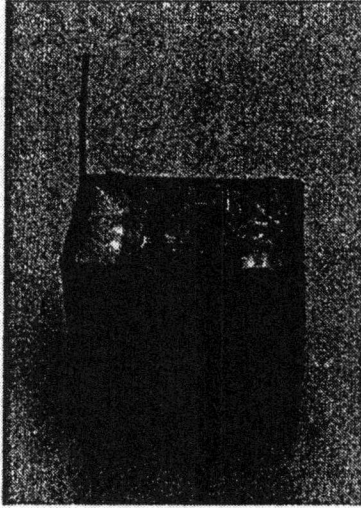

Figure 4.10: The prototype integrated element system in which the capacitor terminal is connected to the transformer. The system is now a three terminal device.

Figure 4.11: Epoxy is used to coat the transformer to create one solid three terminal device.

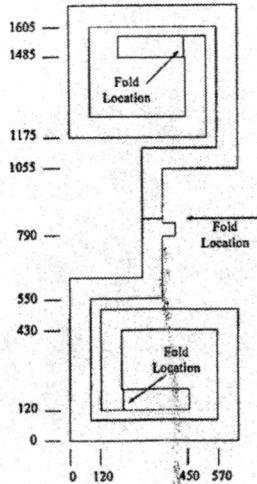

Figure 4.12: The outline of the winding pattern used in the prototype. The dimensions are in mils. The windings are 80 mils wide and there is a separation gap distance of 40 mils when folded.

These transformers were designed such that only one fold is needed to change the planar winding pattern into the desired transformer. The winding patterns associated with the transformer were developed and programmed into OMAX, a software tool used for programming the routing path of drilling, milling machines or in this case a waterjet. The winding pattern selected for the protype design is shown in Fig. 4.12.

The winding pattern is sprayed with glue and a 1 mil thick clear foil is used to insulate the winding. The winding pattern is folded in half above the notch in the center of the winding as shown in Fig. 4.12. The notch is used to make folding easier and provides a larger area to make a solder connection between the center tap of the transformer and one of the pins of the capacitor. The last two legs of the winding pattern were designed to be folded out away from the capacitor to form external leads. A solder connection is made at one electrode and the center post of the transformer and two 22 gauge wires are soldered to the two ends of the transformers in a predefined pinout spacing as shown in Fig. 4.13.

As the base of the capacitor used is filled with potting material, the electrodes and the capacitor windings are some distance from the bottom surface of the actual capacitor winding. The inductance cancellation transformer can therefore be placed directly onto the base of the capacitor and have a sufficient separation distance from the capacitor windings. Isolation distance from the transformer to the bottom of the integrated filter element is set by the thickness of the epoxy layer that is added on the transformer. In the prototype system the epoxy layer increase the height of the capacitor by 68 mils. The other transformers that were designed (with different mutual inductances) require a different spacing in order for them to achieve their optimal performance.

A test of the filter element is made with the test setup shown in Fig. 4.5. The results of this test are shown in Fig. 4.14. These preliminary results show over a 20 dB improvement in impedance for a wide range of frequencies.

The performances of the integrated filter element with the standard capacitor were compared up to 500 MHz (The highest frequency of the network analyzer). At these frequencies the series

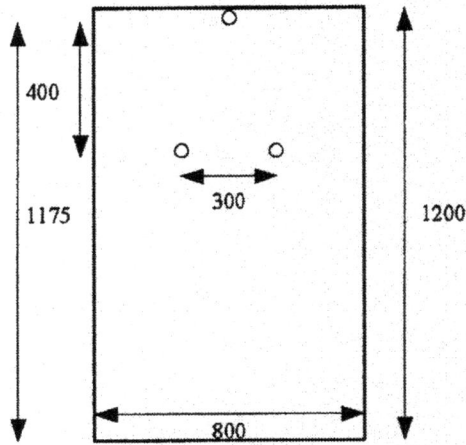

Figure 4.13: The footprint of the integrated filter element. Measurements are in mils.

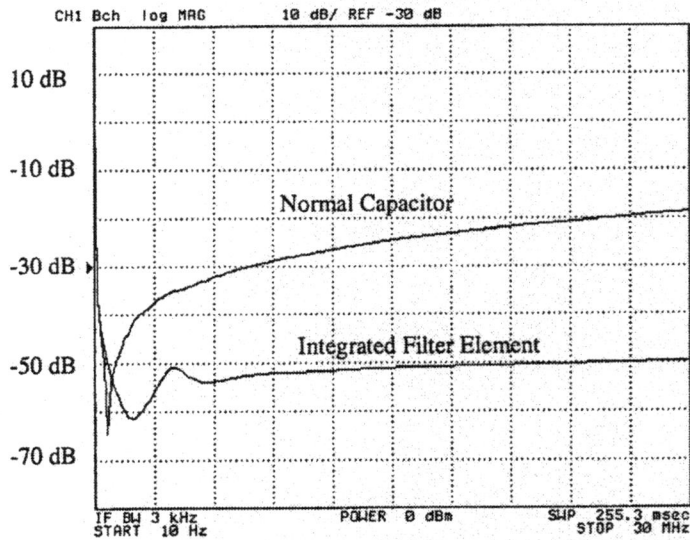

Figure 4.14: Performance results of the device under test using the device shown above. The higher curve shows a typical capacitor. The lower curve shows the prototype integrated filter element with inductance cancellation.

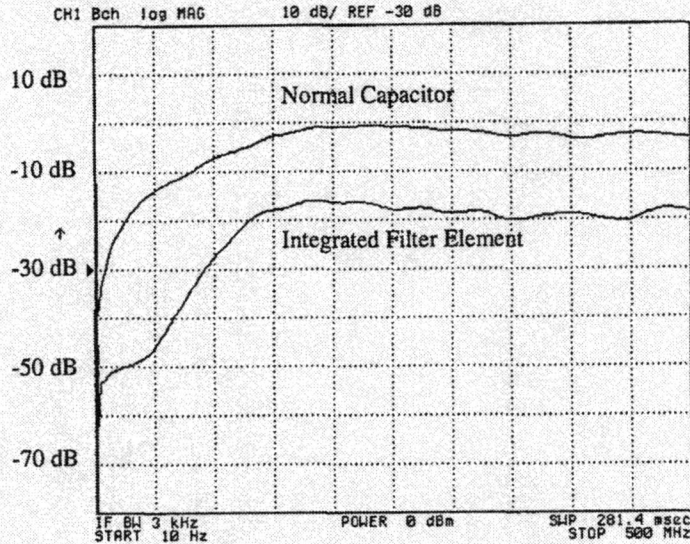

Figure 4.15: Performance results of Fig. 4.14 up to 500 MHz.

inductive impedances introduced by the inductance cancellation technique will be larger than the 50Ω source and load impedances. The output signals from the network analyzer will be a measurement of how much the entire filter network will attenuate high frequency signals. The integrated filter element will show a 30 dB improvement at frequencies less than 50 MHz at higher frequencies the integrated filter element will have 20 dB improvement in attenuation over a normal capacitor as shown in Fig. 4.15.

4.5 Conclusions

Integrated filter elements with inductance cancellation can be constructed using simple automated processes. Two approaches for designing and constructing such elements have been addressed. The first uses a transformer wound about the capacitor body. The second uses a separately constructed transformer that is interconnected with the capacitor. Experimental results demonstrate the efficacy of both construction methods.

Descriptions of end-tapped and center-tapped transformers for the first winding method were shown. The center-tapped transformer will be the smaller of the two transformers, but this detail is less important in the integrated filter element. The comparison between the two should be made after consideration of the manufacturability of the transformers. A method of experimentally determining the number of turns is given along with a method to refine the design. Problems associated with this construction method include introducing material for winding the transformer, attaching tap points to the transformer, and issues with insulation the transformer from the surrounding the voltages.

Means of inexpensively constructing integrated elements using the second approach have been detailed. The transformer can be formed by stamping a planar winding pattern from foil, insulating it, and folding it. It can be packaged together with the capacitor with appropriate insulation or

59

spacing layers. Cost estimates suggest that this is an effective and inexpensive construction method for integrated filter elements. Stamping metal, applying a layer of insulation, folding, encasing in potting material, and making connections to other portions of the capacitor are all simple tasks that can be added to the production of capacitors with only a small increase in cost.

Chapter 5

Inductance Cancellation Circuits with Active Tuning

5.1 Introduction

In some applications it may be desirable to use magnetic materials in a filter with inductance cancellation. For example, this might be done if the cancellation windings are to be integrated as part of another filter component. Implementing accurate and repeatable cancellation of small shunt inductances is an extreme challenge in this case, as the cancellation relies on very precise coupling between the windings, which in turn depends on the properties of the magnetic material. Any mismatch in the coupling (e.g., due to material or manufacturing variations, temperature changes, or mechanical stress or damage) could change the effective shunt inductance and harm the performance of the filter.

To facilitate the use of magnetic materials in filters incorporating inductance cancellation, we propose to develop an adaptive inductance cancellation approach. In this approach, the magnetic coupling of the cancellation windings is placed under closed-loop control, with feedback based on the characteristics of the filter waveforms. This has the potential to achieve maximum filter performance while providing a high tolerance to manufacturing and environmental variations in both the coupled windings and the shunt capacitor.

A major challenge to implementing adaptive inductance cancellation is developing a method to vary the magnetic coupling of the cancellation windings. One approach we will explore for doing this is the use of cross-field magnetic structures [11, 14, 16, 17]. In this approach, a control winding is added to the magnetic circuit in such a way that it does not couple flux with the main windings. The control winding modulates the effective permeability of the magnetic material by driving it into saturation by a controlled amount, thus controlling the inductance and coupling of the main windings.

Even though the majority of applications discussed in this report involve inductance cancellation in capacitors, other elements can benefit from inductance cancellation. In another application the critical inductance that needs to be cancelled may not be as consistent as the equivalent series inductance of a capacitor, either due to part-to-part variations or changes in operating conditions. The use of active tuning will then enable inductance cancellation to be available for these other applications.

This chapter investigates active tuning for inductance cancellation. In this chapter the tunable

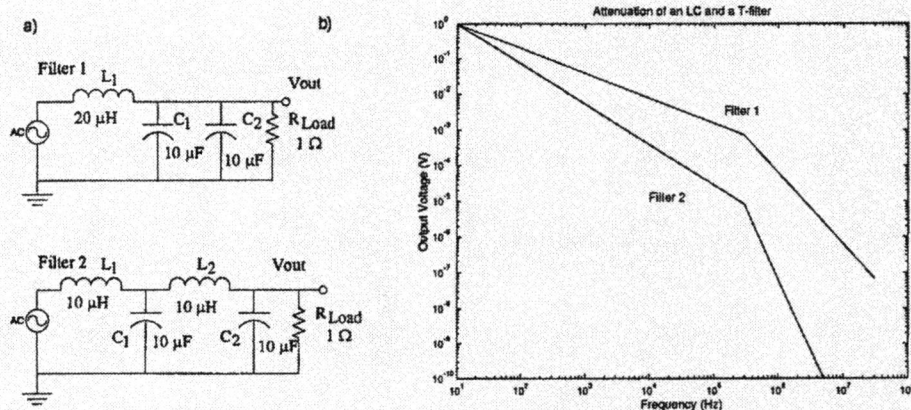

Figure 5.1: The schematics (a) of an ideal L-section filter and a "T-filter". A comparison in attenuation across frequency (b) will show that filter 2, the T-filter, will outperform filter 1, the L-section filter.

element will be separate from the transformer, but both elements may be integrated. Tests results will focus on open loop control of the shunt path inductance of a buck converter. Future work could investigate the closed loop performance of this system.

5.2   Filter Design

Consider the two low-pass filters of Fig. 5.1. In the first filter one large inductor is connected to two capacitors and a load resistor in parallel. In the second filter two inductors, each with half the inductance of the one in the previous example, and the capacitor are connected as a T-Filter which is connected to a second capacitor and the load resistor. The second filter will have more attenuation at high frequencies as shown in Fig. 5.1. The main drawback of the T-filter is the additional cost and volume of the second inductor. The cost and size of the two smaller inductors will typically be higher than the larger inductor but twice as large.

One way to bypass the need of the additional component in a T-filter is to use a transformer as the inductor (a "coupled" inductor). One magnetic structure can supply both inductors in the filter. This filter topology, shown in Fig. 5.2 is almost identical to the "zero-ripple" filter of [5,7–10], illustrated in Fig. 5.3. The key difference between the "zero-ripple" filter and the proposed design is that in the proposed design the shunt path inductance can be controlled and actively tuned to be zero. The "zero-ripple" filter is very sensitive to the parasitics in the capacitor's branch; the transformer has to be designed to have no shunt path inductance. The design of this transformer is very sensitive to variations in construction, materials and operating conditions. With an imperfect transformer and a non-infinite capacitance this "zero-ripple" filter corresponds to the equivalent L-section filter, but with a parasitic shunt-path inductance that is hard to precisely control [8]. The filter style investigated in this chapter, which will be referred to as a modified T-filter, uses active control of inductance cancellation in order to reduce the effect of the shunt path parasitic inductance.

This form of inductance cancellation requires a transformer that is large enough to provide the two large inductances on the T-filter and a method of adjusting the shunt path inductance. Future

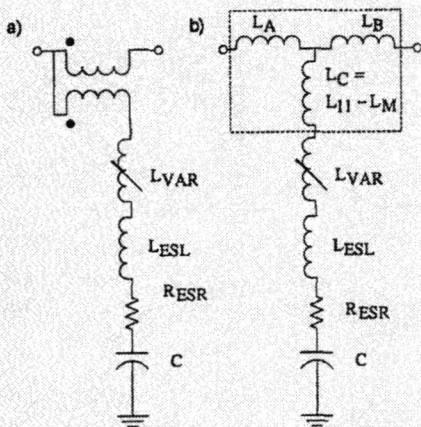

Figure 5.2: Setup for the T-filter with active tuning. The system uses an end-tapped transformer (a) that can be converted to its T-model representation.(b)

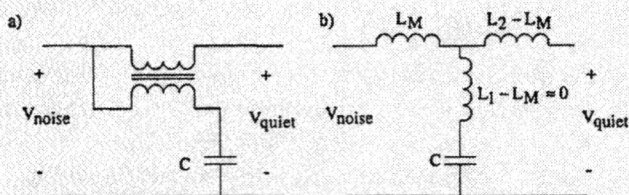

Figure 5.3: The structure of the "zero-ripple" filter (a). Using the T-model of the transformer (b) we can see that the specific design criteria for the transformer is to set L_M equal to L_1.

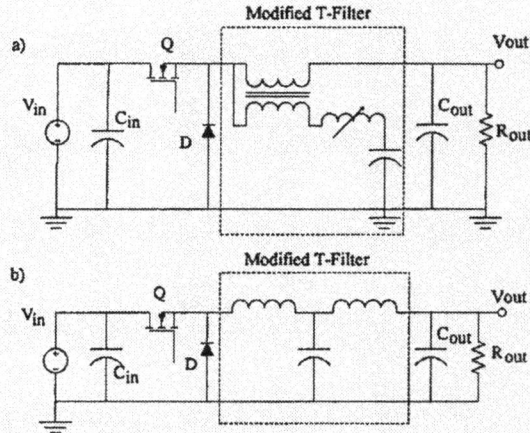

Figure 5.4: A buck converter can use the modified T-filter to replace the buck inductor. The schematic (a) of the filter can be replaced with the T-model (b). The shunt path inductance is set to zero by controlling the variable inductor.

work on this project may investigate methods of achieving both these goals on one structure but in all cases contained in this chapter a separate smaller adjustable inductance will be used in the shunt path.

The inductors in the equivalent T-filter can be integrated into the structure of a converter. For example, the inductor of the buck converter can be replaced by this filter structure. The proposed T-filter can be used instead of using one inductor for the buck converter and an output filter (which may need another inductor). Fig. 5.4 shows a buck converter using the T-filter. The capacitor in parallel with the load will not have inductance cancellation. Although this may improve filtering added inductance in series with the resistive load may not be desired. For this chapter only a comparison involving one element with inductance cancellation will be made.

The transformer may have either an end-tapped or center-tapped connection, as described in Chapter 2. The shunt branch inductance for an end-tapped transformer is $L_{11} - L_M$. The self and mutual inductance of an inductor with a magnetic core are much larger than the parasitic inductances that need to be cancelled. Assuming a high coupling coefficient, the self inductance of the first winding, L_{11}, should just be one or two turns less than the self inductance of the second winding, L_{22}. This will result in a transformer which has both a negative shunt path inductance and a low magnitude. If a center-tapped transformer is used, then the shunt path inductance will be $-L_M$, in order for this to be a viable option the mutual inductance must be small. This will require that either the primary or the secondary winding consist of just one or two turns if the two windings are tightly coupled. The shunt path inductance needs to have a small negative value in order to minimize the size of the variable inductor. Alternately, the two windings of the transformer can be wound on separate legs of a gapped core structure, thereby reducing the coupling

5.2.1 Actively Controlled Inductances

Some kind of active control is needed in order for inductance cancellation to be effective in a circuit that uses an inductor with a magnetic core (and hence has large variable inductance). This method

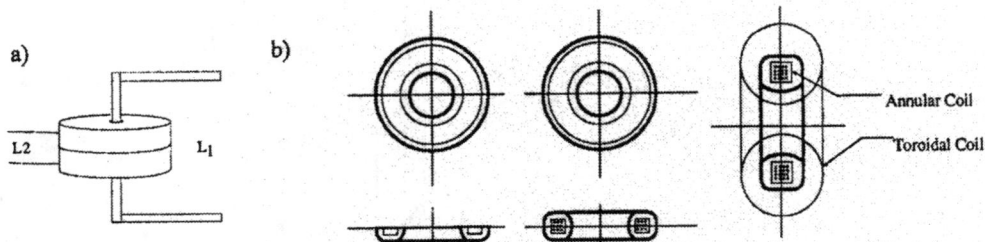

Figure 5.5: Two examples of variable inductors. In a pot core (a) one winding passes through the centerpost hole of a pot core while Another wire is wound in the normal winding window of the core. In a hollow toroid core (b) the annular and toroidal coils are orthogonal.

of control needs to cancel out inductances on the order of tens of nanohenries with enough precision to make the system reliable under a variety of operating conditions and consistent if mass produced. This section addresses the case of actively controlled inductances, or cross field reactors [14], as a method of controlling the shunt path inductance.

A cross field reactor is an inductor with two sets of orthogonally wounds turns [14–17]. Since the windings are orthogonal, ideally the mutual inductance between the windings is zero. For our purposes one winding will form the variable inductance and the other winding will be the control winding. Because the windings are orthogonal ideally the behavior of one winding is independent of the other. However, the control winding is set to partially saturate the core thus influencing the inductance of the other winding. The amount of ampere-turns in the control winding can thus set the variable inductance. A diagram of a cross field reactor using a pot core is shown in Fig. 5.5a.

The maximum value of the variable inductance should be larger than the magnitude of the sum of the shunt path inductance of the transformer and the parasitic inductance of the capacitor. Using the values in Fig. 5.2, the maximum value of the variable inductance must be

$$L_{VAR-max} > -L_C - L_{ESL} \qquad (5.1)$$

The shunt path transformer inductance, L_C, is negative and larger than L_{ESL}. The size of the variable inductor is therefore set by the amount of negative shunt path inductance in the transformer.

The core of the variable inductor is saturated by a control winding. Flux in the core is set by the ampere-turns applied via the control winding, therefore if more turns are used in the winding, then less current needs to be generated to control the inductance. A prototype variable inductor was constructed using a 4C4-40704 pot core with approximately 50 turns of AWG 28 wire in the window area of the core. The variable winding consisted of one turn of AWG 14 wire that passed through the center post hole of the inductor core. The ac inductance of this structure was measured with an impedance analyzer while a DC current was applied to the control windings. A plot of inductance as a function of control current is shown in Fig. 5.6. This plot shows the case when only a small ac current is applied to the variable inductance; in an actual application the ripple current in the filter will pass through the variable inductance and add to the flux in the core. The amount of control current may need to be actively set in order to adjust to different operating conditions.

Testing of the variable inductor in the converter with appreciable ripple currents showed that the ripple current in the main winding can be large enough to change the saturation level of the

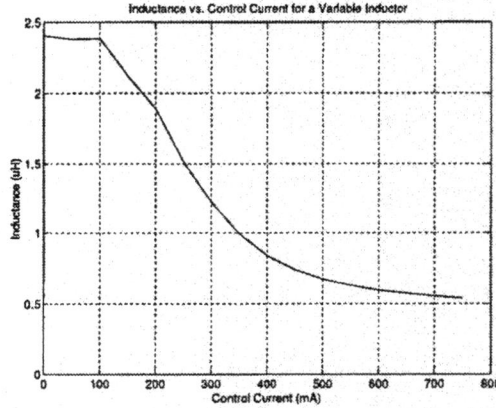

Figure 5.6: The relationship between the variable inductance under small-signal ac condition vs. the control current. For low levels of current (100 mA) the core is not saturated and the inductance is constant. More current into the control windings begins to saturate the core and the inductance drops.

core and cause oscillations (at the switching frequency) in the saturation of the variable inductor. This oscillation will disturb the system every time the current ripple is at its peak. The amount of flux in the core is a vector sum of the control and ripple current. In an appropriately designed system the ripple current should have little impact on the saturation of the core. It is possible to design a custom magnetic structure that has two orthogonal winding paths, in which one path can easily saturate the core and the other path cannot. For the purpose of this chapter a standard magnetic structure will be used that is sufficiently large such that the ac ripple current does not contribute substantially to saturation of the core.

5.2.2 Design Limitations

Two parasitic elements in the design of the modified T-filter: inductance in the shunt path and capacitance bypassing the series path, will both severely degrade performance. The problem of shunt path inductance is well known: the transformer in "zero-ripple" filter designs has traditionally needed to be precisely designed to minimize shunt path inductance[1]. With a magnetic element having a core, deviations from the ideal case (which is not practically achievable) can result in hundreds of nanohenries in the shunt path. The self resonant frequency of the shunt path will be very low if the shunt path inductance is very high. This will lead to reduced high-frequency attenuation. For example, assume that filter 2 in Fig. 5.1a has a shunt path inductance, L_{shunt}, of 200 nH in series with the first capacitor and 15 nH of parasitic inductance in the second capacitor and load, L_{ESL2} and L_{Load}. Above the self resonant frequency of the shunt path, 112 kHz in this example, the filter will have an attenuation approximated by a system of inductances

$$\frac{V_{out}}{V_{in}} = \frac{L_{\text{shunt}}(L_{ESL2} \parallel L_{\text{Load}})}{L_{\text{shunt}}(L_2 + (L_{ESL2} \parallel L_{\text{Load}})) + L_1 L_{\text{shunt}} + L_1 L_2 + L_1(L_{ESL2} \parallel L_{\text{Load}})} \quad (5.2)$$

which in the example is $14.41 \cdot 10^{-6}$. The attenuation of the filter with shunt path inductance

---

[1]Nevertheless, practical fixed designs are not typically able to ever reduce this inductance down to the level of the inductance of the capacitor, as considered elsewhere in this report.

will be approximately constant for frequencies in which the parasitic parallel capacitance of the transformer is negligible.

In traditional "zero-ripple" filters the transformer is designed to have no shunt path inductance, but unavoidable imperfection will mean that in practice a small inductance is present. In the modified T-filter with active control, a variable inductor is used to guarantee that not only does the transformer have no inductance in the shunt path, but that the ESL of the capacitor and any interconnect inductance is also cancelled.

The problem of parasitic capacitance bypassing the series path can be illustrated with an example. A modified T-filter as shown in Fig. 5.2 was designed and evaluated for use in a 100 W 42/14 V dc/dc converter switching at 400 kHz. The transformer windings were chosen such that the shunt path inductance was negative and smaller in magnitude than the variable inductance used in the system. The transformer uses a RM10-315 core made of 3F3 material. The primary winding used 9 turns of AWG 14 to handle the seven amperes of dc current at the output. The secondary used 8 turns of Litz wire (approximately equivalent to AWG 18) to carry the ac currents. The T-model of the transformer has values $L_A = 22.2\mu H$, $L_B = 3.325\mu H$, and $L_C = -2.07\mu H$. The variable inductor is the one used in Fig. 5.6a and the capacitor is $10\mu F$ (106K100CS4) from ITW Paktron.

Testing of this circuit displayed a significant problem in the filter design. The parasitic capacitance associated with the transformer (50 pF in the prototype described above) severely limits the performance. The effect of the parasitic capacitance of the transformer is easily seen in the PSpice simulation results shown in Fig. 5.7. (The actual circuit includes a second capacitor in parallel with the load but this example better demonstrates the problems associated with the parasitic capacitance.) The L-section low pass filter has a resonance at 350 kHz (set by the LC ringing of the capacitor and its ESL) and at higher frequencies no additional attenuation is achieved. The illustrated T-filter also has a resonance above 1 MHz that is due to the presence of the parasitic capacitance. The result is that the performance of both systems is degraded at high frequencies and that the performance improvement expected from use of the T-filter is degraded.

The transfer function of the modified T-filter as shown in Fig. 5.7 is

$$\frac{v_o}{v_i} = \frac{CC_pL_1L_2Rs^4 + (C_pL_1R + C_pL_2R)s^2 + R}{CC_pL_1L_2Rs^4 + CL_1L_2s^3 + (CL_1R + C_pL_1R + C_pL_2R)s^2 + (L_1 + L_2)s + R} \tag{5.3}$$

The zeros of this equation are

$$Zeros = \pm\sqrt{\frac{1}{2CL_1} + \frac{1}{2CL_2} \pm \frac{1}{2C}\sqrt{\frac{1}{L_1^2} + \frac{1}{L_2^2} + \frac{2}{L_1L_2} - \frac{4C}{C_pL_1L_2}}}. \tag{5.4}$$

And as long as the parasitic capacitance is several orders of magnitude smaller than the shunt path capacitance then the location of zeros can be approximated as

$$Zeros \approx \pm\sqrt[4]{\frac{-1}{CC_pL_1L_2}} \tag{5.5}$$

and its complex conjugate. The first resonance of a typical LC low pass filter is at the frequency

$$\omega = \frac{1}{\sqrt{L_{ESL}C}}. \tag{5.6}$$

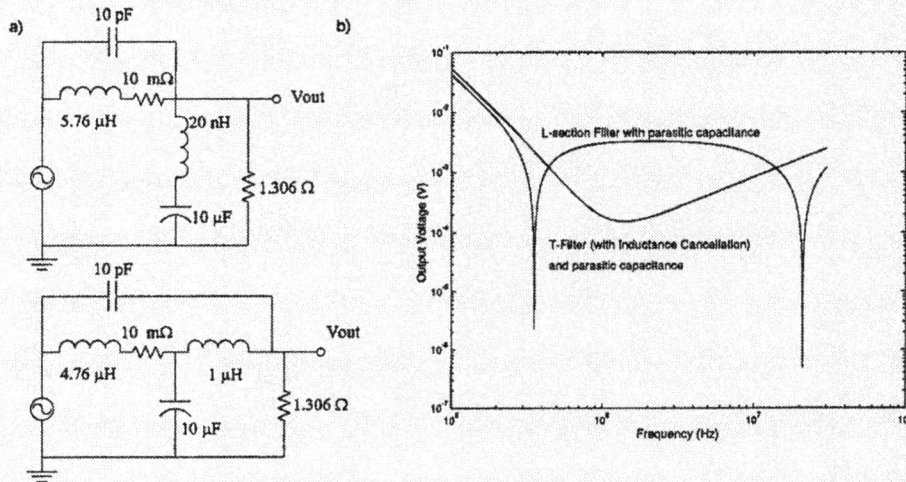

Figure 5.7: A schematic (a) of a T-filter and a corresponding L-section when parasitic capacitance is considered. A simulation (b) comparing the performance of the filters shows a that the modified T-filter outperforms the L-section filter is the zeros associated with the modified T-filter exceed the self resonant frequency of the capacitor in the L-section filter.

To guarantee that that the T-filter will perform better that the L-section filter, an approximate rule of thumb is that the zeros of the improved filter should be at a higher frequency then the resonance of the L-section filter. This limits the parasitic capacitance to

$$C_p < \frac{L_{ESL}^2 C}{L_1 L_2}. \tag{5.7}$$

In the simulation above, C_p is set to 10 pF, the right side of (5.7) is 840 pF. In the experimental setup C_p is 50 pF and the right side of (5.7) is 54.2 pF. This indicates that the experimental system will be no better than an L-section filter, a fact that was experimentally verified.

With this limitation in mind a second system was developed having less parasitic capacitance. The end-tapped transformer consists of five turns of AWG 12 wire for the dc path and four turns of 175/40 Litz wire for the shunt path on a RM10/I-A315 3F3 core. The equivalent T model has the inductances $6.13\mu H$, $1.45\mu H$ and $-0.82\mu H$ for inductances L_A, L_B, and L_C (from Fig. 5.2) respectively. The parasitic capacitance is 9.5 pF and the right side of (5.7) is 450 pF. The variable inductor is a P14/8 A315 3F3 pot core with control windings of 77 turns of AWG 28 wire set through the center post hole of the core. The variable inductance consists of 2 turns of Litz 175/40 wire set through the normal window area of the pot core. The variable inductor with a dc control current between 2 amperes and 0 has a range of inductances from $0.55\mu H$ to $1.26\mu H$. The shunt path capacitor is a $10\mu F$ ITW Paktron capacitor (106K100CS4).

5.3 Comparison of Filter Performance

The performance of a modified T-filter was compared to an equivalent L-section filter and to a "zero-ripple" filter in a buck converter application (e.g., as in Fig. 5.4). A dc/dc converter was

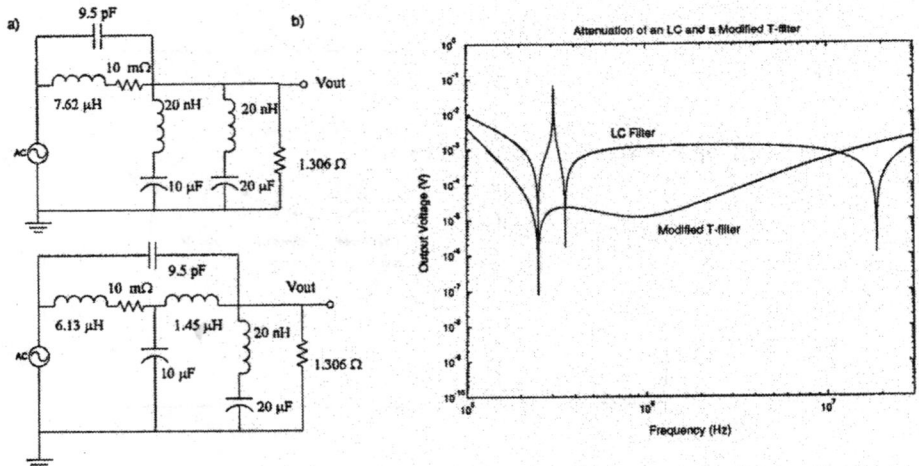

Figure 5.8: A (b) PSpice simulation comparing the attenuation of an (a) L-section filter and a modified T-filter.

built and three different output filters were tested and evaluated.

The dc/dc converter has an input of 42 V and delivers 25 to 100 W of power at 14 V. The switching frequency is 400 kHz. The three filters, shown in Fig. 5.9, all use the same capacitors and the same inductor cores. Fig. 5.8 illustrates the simulated attenuation performance of the L-section and T-filters. A simplified schematic of the converter is shown in Fig. 5.10. The converter operates with an average current inner loop and an outer voltage loop.

The converter was set to its minimum load, 25 W, for testing. Since the converter runs only in continuous conduction mode, the output ripple is independent of the power level. The output voltage ripple was examined with a spectrum analyzer and an oscilloscope; the results are shown in Fig. 5.11 and Fig. 5.12.

The L-section filter uses the same capacitor and the same RM core as used for the transformer in the modified T-filter (RM10/I-A315 3F3) and uses five turns of AWG 14. The measured inductance is $7.62\mu H$. The "zero-ripple" filter transformer was built on the same RM core. The self and mutual inductances are not independent because control of the number of turns, N_1 and N_2, will set and vary the three inductances of the transformer. The transformer winding connecting the input and output is five turns of AWG 14. The winding to the shunt capacitor was varied from four to six turns of 175/40 Litz wire and tested to find which option provides the best performance. The best performance was obtained with a 1:1 turns ratio.

As expected the results of the "zero-ripple" filter are worse than that of an L-section filter. It can be shown that the two filters are very similar by using the T-model representation of the transformer in the "zero-ripple" filter. When real components are considered, as in the filters shown in Fig. 5.9, there will be considerably more shunt path inductance in the "zero-ripple" filter than in an L-section filter, resulting in slightly worse performance even though the "zero-ripple" filter effectively is a T-filter. The comparison in filtering performance is shown in Fig. 5.11 and Fig. 5.12.

The modified T-filter with active cancellation does considerably better than the other two filters. With 400 mA of control current the variable inductor is set such that the shunt path inductance is minimized. With the proper control current the magnitude of the fundamental switching frequency

69

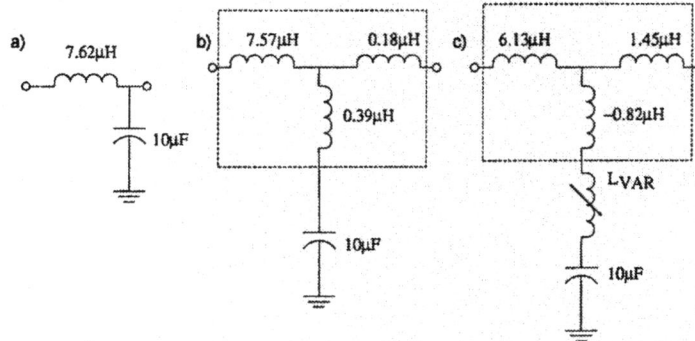

Figure 5.9: The three different filters that will be compared: (a) the L0section filter, (b) the "zero-ripple" filter, and (c) the modified T-filter. All three use the same inductor core and capacitor and the transformer model is replaced with its equivalent T-model

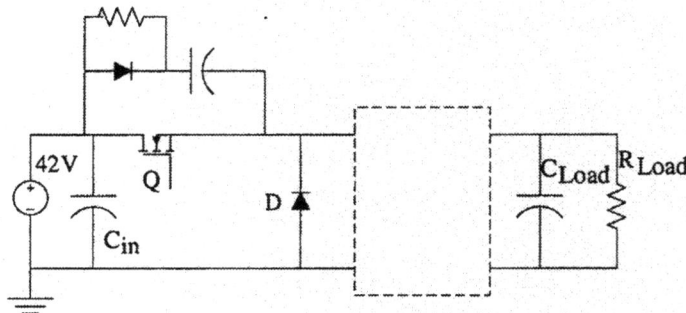

Figure 5.10: A simplified schematic of the experimental dc/dc converter. The box at the output represents either of the three filters shown in Fig. 5.9.(The MOSFET is a SUF75N08-10 and the diode is a 40CPQ080.)

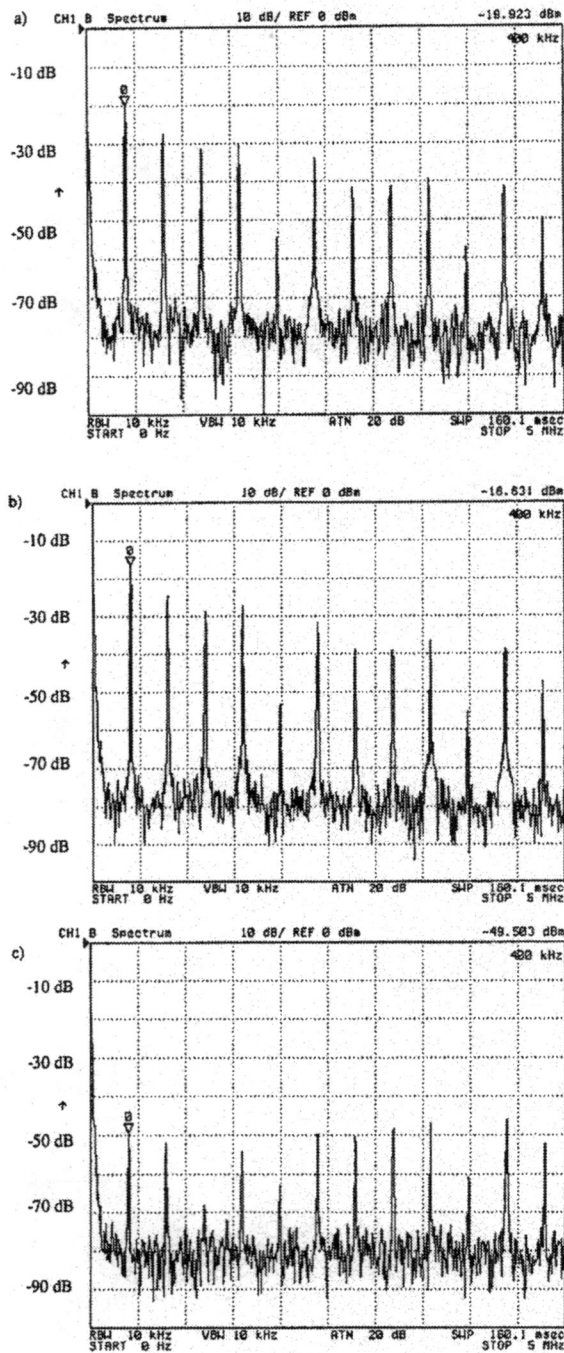

Figure 5.11: The output ripple (frequency domain measurement) of the dc/dc converter using (a) an L-section filter, (b) a "zero-ripple" filter, and (c) a T-filter with actively tuned inductance cancellation.

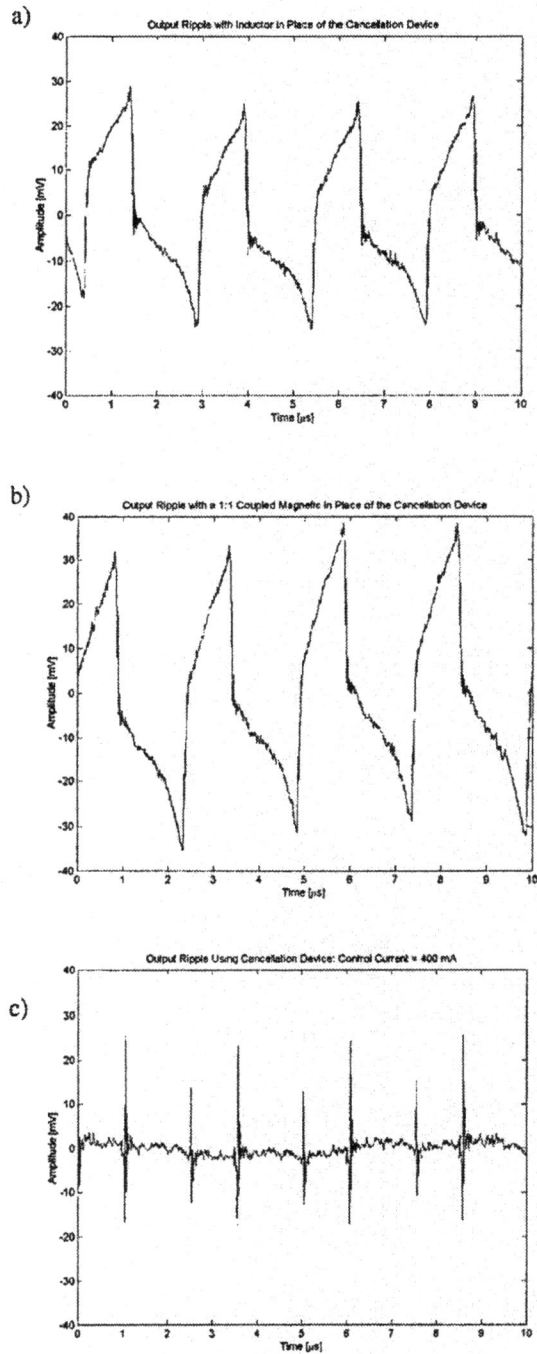

Figure 5.12: The output ripple (time domain measurement) of the dc/dc converter using (a) an L-section filter, (b) a "zero-ripple" filter, and (c) a T-filter with actively tuned inductance cancellation.

72

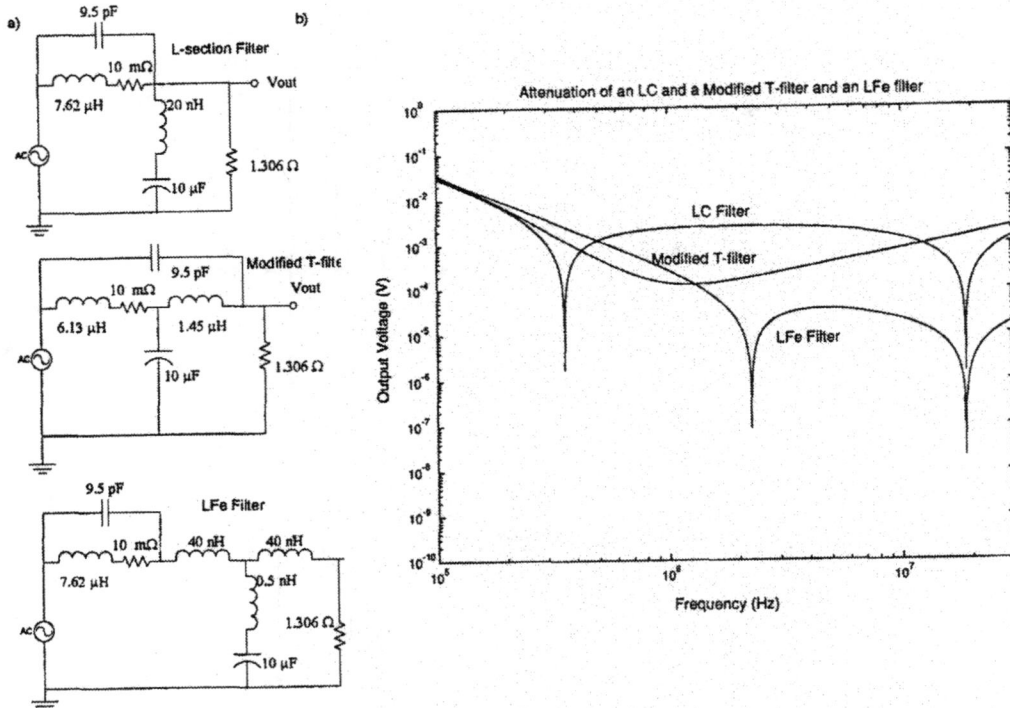

Figure 5.13: The schematics (a) of the an L-section filter, a modified T-filter and a LFe filter. A PSpice comparison of attenuation of the filters. Note that the simulation assumes a 0.5 nH mismatch in the integrated filter element.

is reduced by 30 dB as compared to the equivalent L-section filter. The first four harmonics are significantly reduced, and the next eight are slightly reduced. This performance is expected: as seen in Fig. 5.8 the difference in attenuation between an L-section filter and the modified T-filter is significant for this range of frequencies. For all higher frequencies the difference in attenuation decreases. Above 5 MHz both filters have equivalent performances: at these frequencies the impedance of the transformer or inductor is dominated by the parasitic capacitance.

Another option for an improved filter is to use a standard inductor and a filter element consisting of a capacitor with inductance cancellation as described in the previous chapters. In this chapter this filter will be called an LFe filter. A PSpice simulation of this comparison is shown in Fig. 5.13. A comparison of the LFe to a modified T-filter (with active cancellation) should show results similar to Fig. 5.1 for frequencies below the self-resonance of the transformer or inductor. Therefore to improve the performance of the modified T-filter in relation to the LFe the parasitic capacitance of the transformer or the inductances of the transformer need to be significantly lowered. When the parasitic capacitance dominates the signal throughput of the transformer or inductor then the LFe filter will be better (so long as its parasitic capacitance is negligible).

5.4 Conclusions

This chapter investigates the feasibility of achieving inductance cancellation in a system using a transformer with a magnetic core. A variable inductance formed with a cross-field reactor is used to actively null the shunt-path inductance. This approach is demonstrated using a coupled inductor (transformer) "T-filter", in which the shunt path inductance can be actively cancelled.

Even though the parasitic inductances of capacitors are consistent, there may be other unwanted inductances that are not, whether in terms of part-to-part variation or in terms of variation due to operating conditions. Thus inductance cancellation needs to be actively controlled if applied to systems with an inductance that varies or is not repeatable. Also if an air core inductor is not practical, either because a large inductance needs to be cancelled or excessive current will be flowing through the transformer, then this system can be used.

For the result presented in this chapter the control current into the variable inductor was set by hand. The next step for this project is to close a control loop around the variable inductor such that the output filter will self-adjust to minimize the ripple at the output.

This chapter showed a comparison between a L-section filter, a "zero-ripple" filter and a modified T-filter that all use the same RM core and capacitor. Two limitations shared by the "zero-ripple" filter and a modified T-filter were examined: the presence of shunt path inductance that the modified T-filter can eliminate, and the parasitic capacitance of the transformer that is in the series path. The parasitic capacitance of the transformer will limit the high frequency performance of the filter and it must be kept low. In an experimental setup involving a dc/dc converter the modified T-filter is shown to have 20 to 30 dB improvement in attenuation for the first several harmonics and 5 to 10 db improvement for frequencies up to 5 MHz.

Chapter 6

Multiple Element Inductance Compensation

Conventionally, inductance cancellation windings have only been used to compensate for the equivalent series inductance of a single capacitor. In a filter designed to attenuate both common- and differential-mode signals with multiple capacitors, this significantly increases the number of windings required. The simple EMI filter in Fig. 6.1 contains three capacitors, and the commercial filter in Fig. 6.14 uses six.

The goal of this chapter is to extend the inductance cancellation presented in Chapter 1 by developing a method that allows for the use of a single magnetic winding to compensate for the effects of equivalent series inductances of two capacitors, instead of just one. For many filter topologies,this provides an opportunity to reduce the number of cancellation windings needed, thereby saving precious space and added cost.

6.1 Motivation

To understand why the use of a single magnetic winding to compensate for the inductive parasitics of two capacitors is of particular value in EMI filtering, consider the structure and operation of an EMI filter. Fig. 6.1 shows the basic structure of an EMI filter designed to attenuate both common-mode and differential-mode signals, along with representative source and load networks for performance evaluation. This circuit can be analyzed by separating its common-mode and differential-mode responses and treating these equivalent circuits as if they were independent [37]. The common- and differential-mode equivalent circuits are shown in Fig. 6.2.

Now, if the circuit of Fig. 6.1 is augmented with inductance cancellation coils for each capacitor, the circuit in Fig. 6.4 is generated. In this new figure, the differential capacitor C_X is fitted with two inductance cancellation coils instead of only one to preserve circuit symmetry. Past work [38] has shown this to be as effective as a single coil, and Fig. 6.3 shows a photograph of this where the inductance cancellation windings are fabricated on a PCB.

It is desirable to implement the cancellation windings in a balanced fashion to avoid inserting an unbalanced circuit element within the otherwise well-balanced system. Without balancing the series inductances on both sides of the capacitor, a cross coupling between the differential and common-mode signal sources would result. By avoiding this coupling, the common- and differential-mode circuit equivalents remain straightforward, as illustrated in Figs. 6.5(a) and 6.5(b).

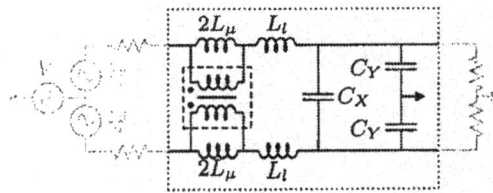

Figure 6.1: Simple EMI filter circuit shown with representative source and load networks for performance evaluation. Some parasitic elements (such as capacitor equivalent series inductance) are not shown explicitly.

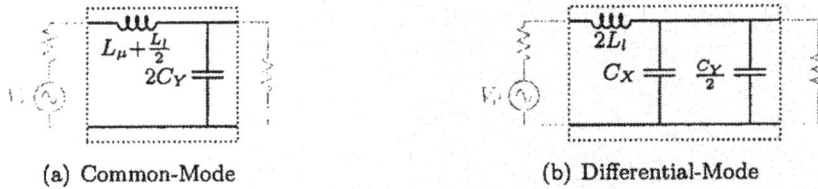

(a) Common-Mode

(b) Differential-Mode

Figure 6.2: Models for the simple EMI Filter circuit of Fig. 6.1, decomposed into common- and differential-mode portions.

Figure 6.3: Test circuit with balanced inductance cancellation windings implemented in the printed circuit board. Performance of this filter has been previously shown [38].

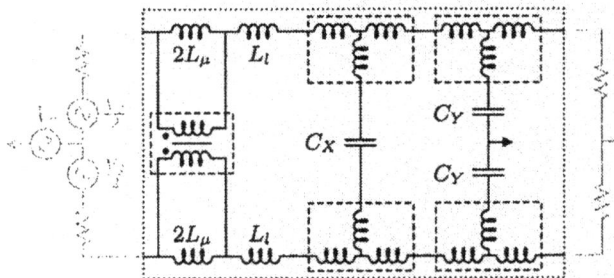

Figure 6.4: Simple EMI Filter circuit from Fig. 6.1 with balanced inductance cancellation of each capacitor.

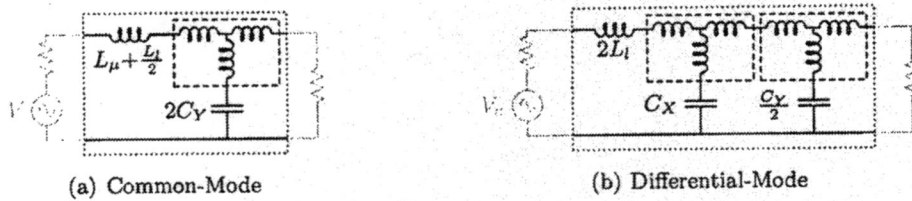

(a) Common-Mode   (b) Differential-Mode

Figure 6.5: Simple EMI Filter circuit with balanced inductance cancellation of each capacitor, decomposed into common- and differential-modes.

As shown in Fig. 6.4, the construction of an EMI filter with full, balanced inductance cancellation would require four magnetically coupled windings when constructed using the previously established method. These windings occupy additional space within the filter, and if placed in close proximity may exhibit secondary effects from magnetic coupling, complicating the design. Given these limitations, it would be a considerable improvement if the number of required windings could be reduced by utilizing a single winding to provide appropriate inductance compensation for two capacitors.

6.2 Implementation

To show experimentally that the use of a single inductance cancellation coil for two capacitors is feasible, a simple test filter was created with a planar winding mounted with EMI filter capacitors inside a shielded enclosure. Fig. 6.6 shows the filter along with the two Panasonic ECK-ATS472ME6 4700pF Y2 class ceramic capacitors used. This test filter does not directly examine common- and differential-mode testing, however it does provide a straightforward example how a single coil can support the compensation of inductance for two capacitors. A dimensioned line-art drawing of the coil, which was cut using an OMAX abrasive-jet cutter from a single piece of 1mm thick copper, is shown in Fig. 6.7. Based on simulation results from FastHenry [39], the coil itself has a maximum series inductance of 393.0nH, and a maximum equivalent shunt-path inductance of -63.2nH when used for single element inductance cancellation (in the magnetic winding T model). It should be noted that this coil was intentionally designed to be far over-sized for the amount of cancellation required; this was to allow for maximum flexibility in testing.

The procedure outlined here was developed for tuning the filter response of the two capacitors, and is one way a high performance filter response can be determined. Initially, the connection of capacitor C_1 is tuned to optimally cancel its parasitic inductance. This can be done by adjusting the connection point of the capacitor on the winding while observing the filter attenuation (e.g. with a network analyzer), and/or using methods associated with previously described techniques in [40]. Once its optimal position is found, the position of the capacitor is fixed. Following this, the connection of capacitor C_2 is tuned (with capacitor C_1 in place) to find an optimal filter response. This gives one possible combination of capacitor locations on the coupled winding that results in a high performance filter characteristic.

Experimental results for this test system are shown in Fig. 6.8, with data taken from an Agilent 4395A network analyzer which provides 50Ω source and load impedances. Insertion gain measurements were made in accordance with those used to evaluate inductance cancellation performance in [38, 40] to allow for direct performance comparison. When tuning the response with only C_1,

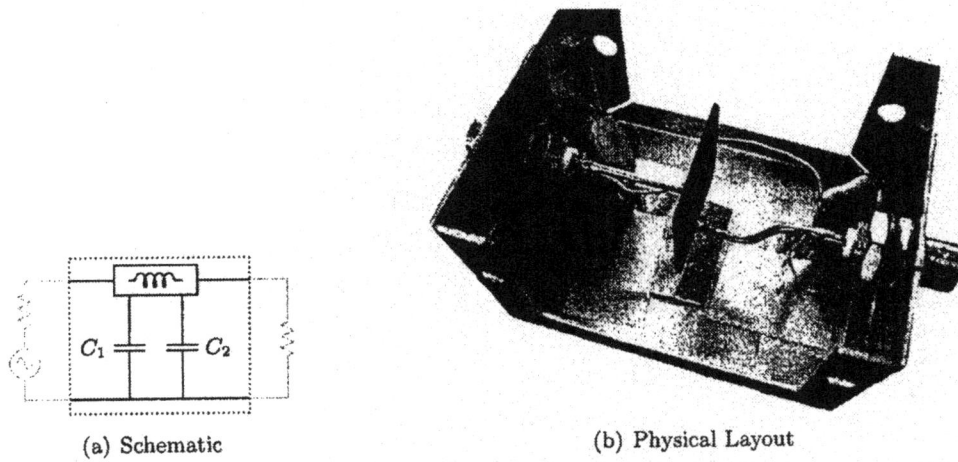

(a) Schematic   (b) Physical Layout

Figure 6.6: Test filter for inductance compensation of two Panasonic ECK-ATS472ME6 4700pF ceramic capacitors using a single magnetic winding.

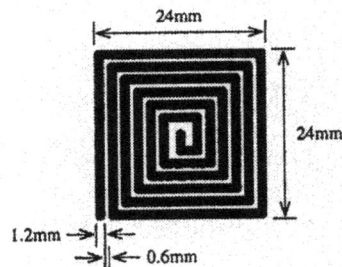

Figure 6.7: Illustration of the planar winding used in the test filters of Section 6.1, fabricated from 1mm thick copper. The total series-path inductance based on simulation is 393.0nH, and the maximum equivalent shunt-path inductance for a single element is -63.2nH (in the magnetic winding T model).

78

two measurements were taken for comparison: one with the capacitor connected directly at the input (source-side) terminal providing no cancellation, and one where the capacitor was connected to the cancellation coil at a location where the output response was optimal. The same approach was taken when tuning the response for the combination of C_1 and C_2: C_2 was connected either directly at the filter output (load-side) terminal or at a position optimizing the filter response with both capacitors.

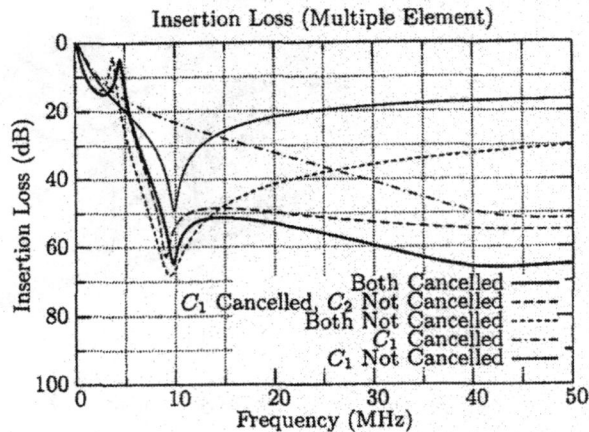

Figure 6.8: Measured results from the test filter in Fig. 6.6 showing the performance of multiple-element inductance compensation.

The characterization results of the filter attenuation performance clearly show a dramatic improvement (as much as 35dB at high frequency) from the case where no compensation is provided (*Both Not Cancelled*) to the case where inductance compensation is provided for both capacitors (*Both Cancelled*). These results demonstrate that a single coupled magnetic winding can be used to provide inductance compensation for two capacitors, with dramatic performance improvement at high frequencies.

6.3 Coupling of Multiple Windings

When physically placing multiple magnetic windings in close proximity, linked magnetic flux between the windings can affect the predicted performance in various ways [41]. Thus, the implementation of multiple cancellation windings in a single filter may affect the inductance cancellation and filter performance. Here the effects of mutual coupling are explored when two coils are used to provide balanced inductance cancellation for both common- and differential-mode capacitors.

Two additional filters (using the same windings shown in Fig. 6.7) were created to test two coil configurations having different magnetic coupling directions. In addition to a pair of line-to-ground (Y) capacitors (Panasonic ECK-ATS472ME6) for common-mode filtering, these test filters incorporate a Rubycon 250MMCA334KUV class X2 line-to-line capacitor for differential-mode filtering. Fig. 6.9 is a photo of one of the filters, and shows its internal layout. Figs. 6.10 and 6.11 show the filter configurations and illustrate the difference between the two winding orientations.

Windings placed in the *same direction* each throw flux in a way which opposes the flux of its paired winding for common-mode currents, reducing each winding's effective inductance. In the

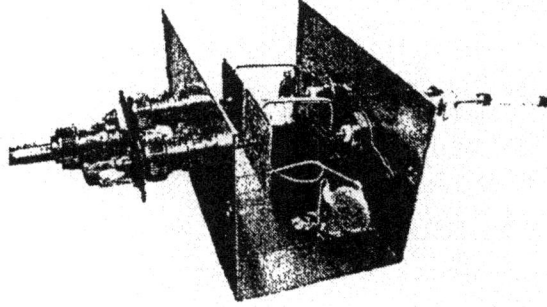

Figure 6.9: Filter for investigation of common- and differential-mode coupling between inductance compensation windings.

case of the windings oriented in the *opposite direction*, the flux from each winding is reinforced by the other for common-mode currents, providing a coupling direction like that of a common-mode choke, and increasing each winding's effective inductance.

(a) Same Direction

(b) Opposite Direction

Figure 6.10: Two orientations of coupled inductance compensation coils. The coils are of the type shown in Fig. 6.7. C_{Y1} and C_{Y2} are Panasonic ECK-ATS472ME6, C_{X1} is a Rubycon 250MMCA334KUV. The two circuits only differ with respect to mutual coupling among the coils.

The tuning procedure used here is similar to the one used in the two-capacitor case in Section 6.2. Initially, the connections of capacitors C_{Y1} and C_{Y2} are tuned simultaneously in the common-mode case to compensate for their parasitic inductances (while retaining a balanced configuration). Once the optimal positions are found, the positions of the capacitors are fixed. Following this, the capacitor C_{X1} is tuned in the differential-mode case by moving its connections on both coils symmetrically to find an optimal output response. Tuning is carried out in this order because ideally the addition of the differential-mode capacitor does not affect the common-mode response, while the reverse would not necessarily be true.

The measurement setup for the common- and differential-mode filter insertion-loss performance is taken from [42], with signal generation and measurement performed by the same Agilent 4395A Network Analyzer as in Section 6.2, with Mini-Circuits 180° power splitters (models ZSCJ-2-1 and ZSCJ-2-2) for dividing its output into differential signals, and custom-made common-mode splitters.

In both winding configurations the target frequency for optimization was 30MHz, with measurements shown up to 40MHz. The two orientations possess similar optimized filtration performance,

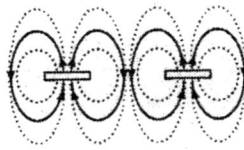

(a) Same Direction   (b) Opposite Direction

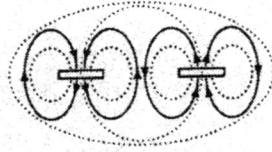

Figure 6.11: Flux patterns for common-mode operation of the two magnetic winding configurations of Fig. 6.10. Windings oriented in the *same direction* generate flux in a way which opposes the flux of the paired winding for common-mode currents. Windings oriented in the *opposite direction* generate fluxes which reinforce each other, providing a coupling direction like that of a common-mode choke.

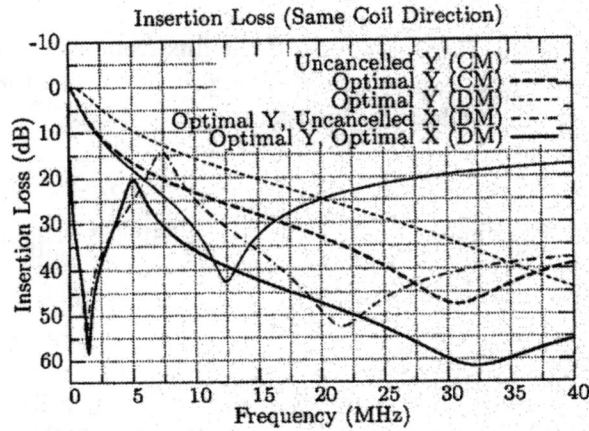

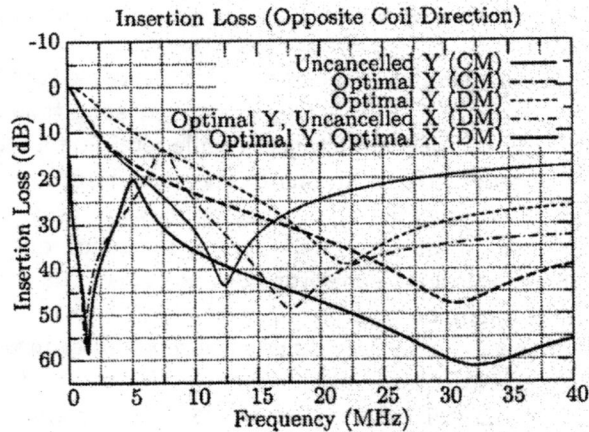

Figure 6.12: Measured results from the coupled inductance compensation winding orientations of Fig 6.10, including both common-mode (CM) and differential-mode (DM) measurements.

81

seen in the thicker traces of Fig. 6.12. The thinner traces in show additional measurements from intermediate steps in the tuning process.

The results show that in both winding orientations an equivalent inductance compensation improvement can be achieved for both the common- and differential-modes. This allows the orientation of the windings to be selected based on other factors (e.g. based on magnetic coupling with more dominant circuit parasitics). While the winding orientation does not influence the final optimized response in these filters, how each winding orientation achieves this optimum is slightly different. In Fig. 6.13 the connection locations for the filter capacitors are shown, corresponding to the optimal common- and differential-mode filter response from Fig. 6.12.

(a) Same Direction   (b) Opposite Direction

Figure 6.13: Connection locations of capacitors corresponding to the results in Fig. 6.12. Only one winding of each pair is shown, the connections made to the other winding are symmetric. X represents the connection location of C_{X1}, Y represents the corresponding C_{Yn} connection location for that winding, I is connected to the input of the filter, and O represents the connection to the filter output.

Due to the coupling in the common-mode, the connection for the C_{Yn} capacitor was closer to an end terminal on the winding in the *opposite direction* orientation than in the *same direction* orientation. Effectively, in the common-mode, the *opposite direction* orientation has a marginally higher inductance-per-turn than the *same direction* orientation, and thus requires a slightly reduced number of turns to achieve the same performance.

Even with the the windings in close proximity, the effects of magnetic coupling on the inductance compensation are minimal. In more extreme cases where the coupling is significantly higher, the observed effects may become more pronounced. Even in this case, however, an equivalent performance should be achievable given properly sized windings.

6.4   Application to Commercial EMI Filter

Having shown in the previous sections that a single inductance cancellation winding can be used with two capacitors to improve filtration performance, and that the coupling orientation of multiple windings in a single filter does not adversely affect potential inductance compensation, the use of multiple element inductance compensation in the context of common- and differential-mode EMI filter is examined. A commercially-available filter is used as a starting point.

Figs. 6.14(a) and 6.14(b) show the schematic and physical views of the filter, which is rated for up to 250 volts and 25 amps of 50/60Hz alternating current. The large (15μH) series inductors L_{11} and L_{12} are particularly bulky, heavy, and expensive components of the commercial filter, and it would be desirable to eliminate them provided that filter performance is preserved. The series inductors were removed to provide working space for installing the inductance cancellation windings, and to provide an opportunity to offset their removal through use of the much smaller cancellation windings. Figs. 6.15(a) and 6.15(b) show the modified schematic and physical layout of the filter with the inductance cancellation windings installed. Additionally, Fig. 6.16 shows the folded design of the inductance cancellation coil used in this filter. As with the previous coil in Fig. 6.7, Fig 6.16 was cut with an abrasive-jet cutter, using 2mm thick copper for enhanced current carrying capacity. The flat winding structure is folded at the center of its longest side to form a square one-piece two-layer winding with Mylar tape used as insulation between the layers. Based on simulation results, the coil is estimated to have a series inductance of 288.3nH, and a maximum equivalent shunt-path inductance of -81.2nH when used for single element inductance cancellation (in the magnetic winding T model). As in the previous test filters, the coil is purposefully over-designed for the required inductance cancellation to allow for additional design flexibility and testing.

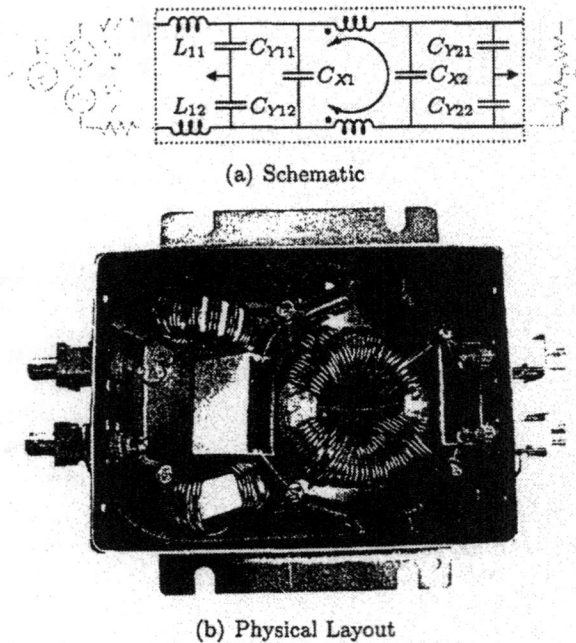

(a) Schematic

(b) Physical Layout

Figure 6.14: Original Commercial EMI Filter. L_{11}, L_{12} are 15μH wound toroidal inductors, C_{Y11} and C_{Y12} are Rifa PME-271 47nF film capacitors, C_{X1} and C_{X2} are Vishay Roederstein F1772-522-2030 2.2μF film capacitors, C_{Y21} and C_{Y22} are 15nF ceramic capacitors, and the common-mode choke has measured leakage inductances of 30.2μH and a magnetizing inductance of 4.45mH.

Common- and differential-mode measurements were taken of the unmodified filter, as well as an intermediate step before the inductance cancellation windings were installed. In this intermediate step, the large inductors L_{11} and L_{12} were removed and straight, solid 14ga wire was installed in their place. This configuration, referred to here as *Without Series Inductor*, was used as a baseline comparison for improvements based on inductance cancellation.

The tuning procedure outlined here is the same as the one used in Section 6.3, and was developed for tuning the filter response due to the common- and differential-mode capacitors. Initially, the connections of common-mode capacitors C_{Y1} and C_{Y2} are tuned simultaneously to compensate (in

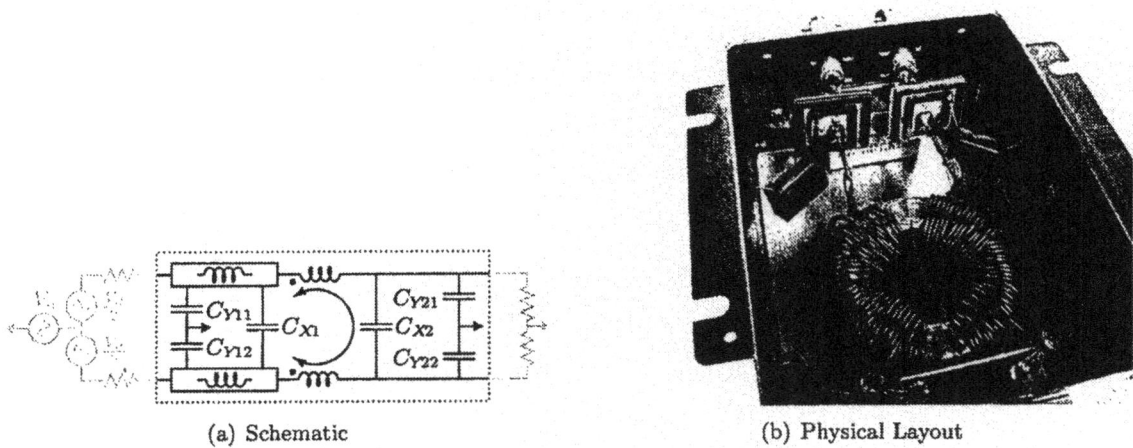

(a) Schematic

(b) Physical Layout

Figure 6.15: Modified version of the EMI filter in Fig. 6.14 with L_{11} and L_{12} removed, and two inductance compensation windings installed.

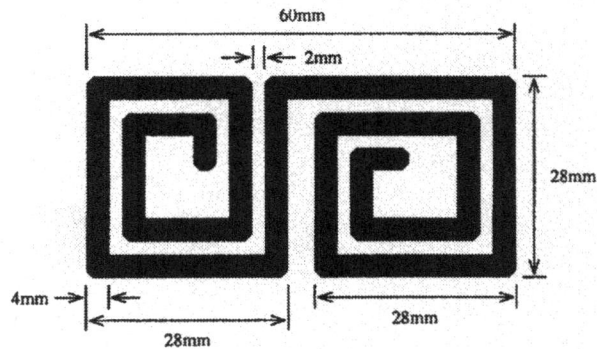

Figure 6.16: Illustration of folded winding used for inductance compensation in the EMI filter of Section 6.4, fabricated from 2mm thick copper. When folded, the total series inductance is 288.3nH, and the maximum equivalent shunt-path inductance for a single element is -81.2nH (in the magnetic winding T model).

84

Table 6.1: Published common- and differential-mode 50Ω circuit insertion loss specifications for the commercially-available EMI filter considered in Section 6.4. All measurements are listed in dB.

	Frequency (MHz)								
	0.01	0.03	0.05	0.1	0.5	1.0	5.0	10.0	30.0
Common-Mode	2	14	22	36	75	75	70	70	48
Differential-Mode	14	14	17	42	75	75	70	70	50

a symmetric fashion) for their parasitic inductances. Once the optimal positions are found, the capacitors are permanently attached to their respective windings. Following this, the differential-mode capacitor C_{X1} is tuned by moving its connections on both coils symmetrically to find an optimal output response.

This order of tuning makes sense: the common- and differential-mode capacitors do not impact system performance in the same way. In Fig. 6.5(a) it can be seen that the common-mode equivalent circuit is not influenced by the differential-mode capacitance (or the inductance cancellation, other than through the fixed series inductance introduced by the winding); the common-mode filtration operates as if the differential-mode capacitor were an open circuit. However, the differential-mode filtration is dependent on the common-mode capacitance and inductance cancellation. This means that if the inductance compensation for the common-mode capacitance is optimized first, the inductance compensation for the differential-mode capacitor can be tuned subsequently without influencing the common-mode performance.

The results of the completed tuning are shown in Fig. 6.17 along with the stock and baseline filter configurations. It should be noted that tuning of both the common- and differential-modes is based on compromises between high and low frequency performance. This particular "optimal" output response chosen here may not be the highest achievable performance for a particular range of frequencies of interest.

The results of incorporating the inductance cancellation coils reveal a dramatic improvement in the filtration performance for both the common- and differential-mode responses over the baseline (stock filter with L_{11} and L_{12} removed, labeled as *without series inductor*). The common-mode shows improvement across its full range, and the differential-mode shows substantial improvement over its full range except for the small resonance around 2MHz. (This small resonance is caused in part by the capacitor-inductor-capacitor π-section formed with the two capacitors attached to the inductance compensation winding in the differential-mode, and can be reduced by fabricating a winding with lower series inductance than the over-sized one used here.)

The common-mode performance with inductance compensation is somewhat worse than the stock filter by approximately 10dB, while the differential-mode performance is very comparable, even without L_{11} and L_{12}. More importantly, the performance with the inductance cancellation windings exceeded the commercially-published performance specification of the stock filter, shown in Table 6.1, without requiring the large, expensive series inductors of the stock filter.

The results from this commercial EMI filter, as well as those from the test filters in Section 6.1, show clearly that a single magnetically coupled winding can provide effective inductance compensation for two capacitors. Moreover, it is demonstrated that the performance of a commercial filter design can be preserved at lower component weight and cost through use of the proposed approach. It is anticipated that further substantial design improvements could be achieved in a filter expressly designed to take advantage of the inductance compensation method proposed here.

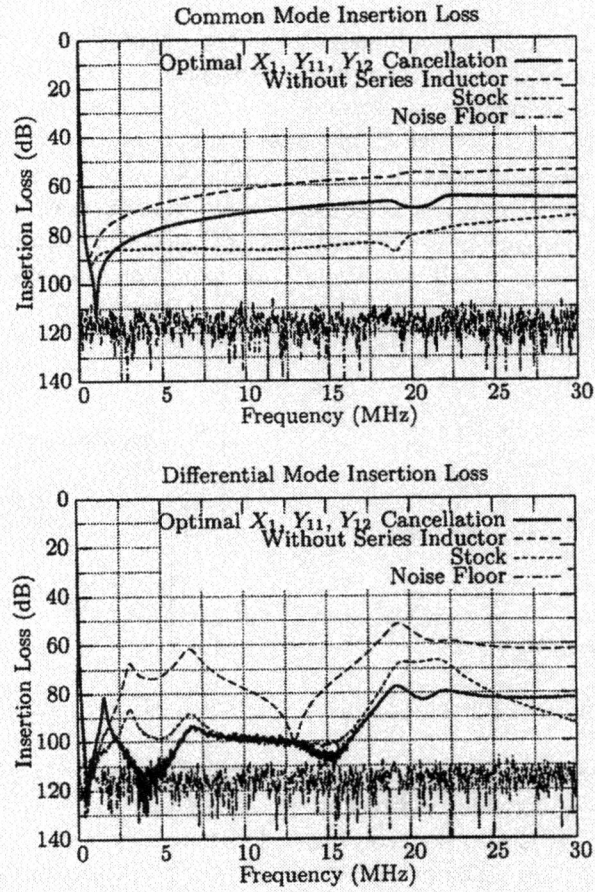

Figure 6.17: Performance comparison of the commercially-available EMI filter in Fig. 6.14 and the modified version in Fig. 6.15, showing performance both without and with inductance compensation.

6.5 Analytic Formulations

In this section an analytical basis is sought for the proposed method of compensating for the inductance of two capacitors using a single coupled magnetic winding. It is derived from an extension of the methods used to analyze single-capacitor inductance cancellation techniques. The predictions of this method are then compared to measured results to illustrate its usability.

6.5.1 Extended Cantilever Model

Analysis of inductance cancellation windings with a single capacitor is relatively straightforward since a two-port transformer model of the windings is used, which has only three independent terms. The number of independent terms needed to completely describe coupled magnetics with n terminals is given by $n(n+1)/2$ [43], which grows as the square of n.

Adding to the complexity is the fact that many models for multiple winding transformers either do not adequately model the complete transformer behavior, or have poor correlation and numeric conditioning to attempted measurements of model parameters from terminal characteristics [44,45]. One model that is effective, and well conditioned for experimental parameter extraction, is the Extended Cantilever Model [43,44].

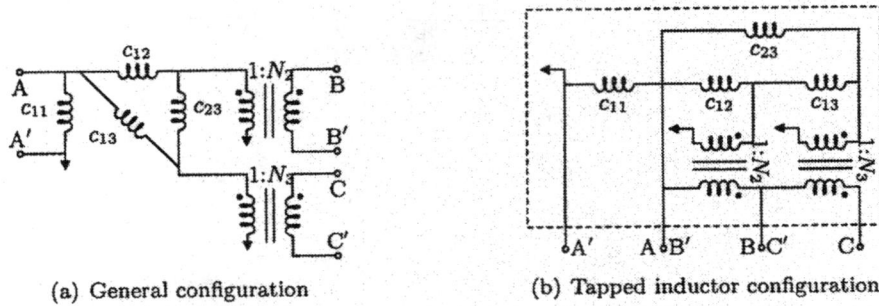

(a) General configuration

(b) Tapped inductor configuration

Figure 6.18: Three-port extended cantilever models.

The Extended Cantilever Model of a coupled system yields an equivalent circuit with directly measurable parameters and provides a direct mapping between circuit parameters and the inductance matrix parameters. It is also well conditioned numerically when dealing with small leakage fluxes or high coupling factors. The extended cantilever circuit model for a three-port system is shown in Fig. 6.18(a), with circuit parameters related to impedance matrix parameters as follows:

$$Z = sL \tag{6.1}$$

$$B = \begin{bmatrix} Z_{11} Z_{12} Z_{13} \\ Z_{21} Z_{22} Z_{23} \\ Z_{31} Z_{32} Z_{33} \end{bmatrix}^{-1} \tag{6.2}$$

$$N_k = \frac{Z_{1k}}{Z_{11}} \tag{6.3}$$

$$c_{11} = Z_{11} \tag{6.4}$$

$$c_{ij} = -\frac{1}{N_i N_j b_{ij}} \qquad (6.5)$$

where b_{ij} is the $(i,j)^{\text{th}}$ element of B.

It should be clarified that the notation used to indicate the impedance matrix Z is representative of only the reactive component of Z due to the equivalent inductance; the extended cantilever model in [43] is formulated only with consideration to inductances. An extension which allows for full Laplace domain circuit elements can be found in [46, 47], however in the idealized case considered here, parasitic resistances and capacitances are assumed to be negligible.

6.5.2 Three-Port Analysis

Fig. 6.18(b) shows the application of the extended cantilever model to a center-tapped winding with two tap points. Fig. 6.19 shows additional circuit connections used for finding the system transfer function. The full transfer function for the system is given in Appendix D, and truncated versions are utilized in this section where appropriate.

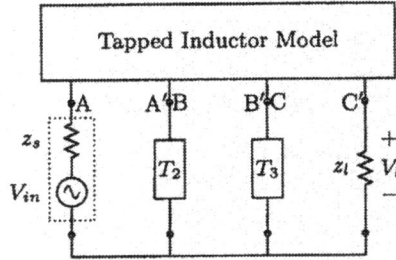

Figure 6.19: Shared-terminal three-port circuit for use with the Extended Cantilever Model tapped-inductor configuration of Fig. 6.18(b). T_2 and T_3 represent the (inductive) high frequency impedances of the capacitors.

By analogy to the case of inductance cancellation for a single capacitor, we desire to find conditions that drive the transfer function from the input source to the output voltage to zero (or close to zero). In finding where the transfer function goes to zero, conditions must be found where both the numerator becomes zero, and the denominator remains finite and non-zero. Starting from the numerator of the full transfer function in (D.1) from Appendix D, setting it equal to zero, refactoring, and dividing by the non-zero value of z_l, a condition is found in which a zero in the numerator can be generated:

$$\begin{aligned} 0 = {} & Z_{13}\left(Z_{12} + Z_{23} + Z_{13} + Z_{22}\right) \\ & - \left(T_2 - Z_{12} - Z_{13}\right)\left(T_3 - Z_{13} - Z_{23}\right) \end{aligned} \qquad (6.6)$$

This result, considered by itself, provides a number of terms that can be adjusted to satisfy the equality. However, in the case of common-mode and differential-mode filtering there are additional constraints that must be considered.

6.5.3 Common- and Differential-Mode Optimization

As described previously in Sections 6.3 and 6.4, in an EMI filter the common- and differential-mode capacitors do not impact system performance in the same way: the common-mode filtration operates as if the differential-mode capacitor were an open circuit, while the differential-mode filtration is dependent on the common-mode capacitance and its inductance cancellation. To find the optimal cancellation for the common-mode capacitor, the transfer function in (D.1) is considered at the limit where $T_3 \to \infty$ (the differential-mode capacitance is a virtual open circuit for common-mode signals).

$$H(s) = \frac{(T_2 - Z_{13} - Z_{12}) z_l}{\cdots} = 0 \tag{6.7}$$

From this result, it is shown that if $T_2 = Z_{12} + Z_{13}$, then full cancellation in the common-mode can be achieved. With the common-mode cancellation constraint met, the result is then inserted back into the original transfer function in (D.1) to find the constraint placed on the differential-mode compensation:

$$H(s) = \frac{Z_{13} (Z_{12} + Z_{23} + Z_{13} + Z_{22})}{\cdots} = 0 \tag{6.8}$$

The numerator of this result contains no terms of T_3 in which to tune in comparison to the terms of the impedance matrix. Additionally, the terms of the impedance matrix in a cylindrically or concentrically wound coil configuration are positive, preventing simple geometries from creating a zero in the transfer function. If differential-mode compensation is to be achieved, this result seems to provide no opportunity for the transfer function to become zero, save for the possibility of making $Z_{13} = 0$. In the case where Z_{13} can be made zero, the transfer function denominator would remain finite and non-zero, representing a possible condition to generate a zero for the transfer function, if the structure can be arranged to provide it.

Another consideration may apply in this case. In past work [40] it was shown that depending on the frequency range of interest, filter performance, even with imperfect cancellation, may be perfectly adequate for practical purposes. With imperfect cancellation, a new term Δ_2 can be defined to be the effective residual shunt-path impedance of the capacitor. More specifically, $\Delta_2 = T_2 - (Z_{12} + Z_{13})$. If this is substituted into the general condition in (6.6), and with the resulting equation rearranged, (6.9) results. This provides a relation where Z_{13} is not explicitly required to be zero for the transfer function to become zero.

$$0 = Z_{13} (T_2 + Z_{22} + Z_{23}) - \Delta_2 (T_3 - Z_{23}) \tag{6.9}$$

Hence, one may gain good performance in both common-mode and differential-mode by realizing substantial (but not perfect) cancellation in common-mode to benefit differential-mode performance.

6.5.4 Simulation and Model Validation

To validate the model and transfer function analysis, the common- and differential-mode filters constructed in Section 6.3 are used as the basis for simulation, excluding the coupling between the two coils. Each of the measurements presented in the section is simulated here using the model developed, to allow comparison to the experimental results.

To simulate both common- and differential-mode responses, the equivalent circuit models for each mode are constructed. These equivalent circuit models, shown in Fig. 6.20, include the inductance compensation windings, as well as the equivalent series resistance and equivalent series inductance of each capacitor. This allows for the direct use of (D.1) from Appendix D, the transfer function of the circuit in Fig. 6.19.

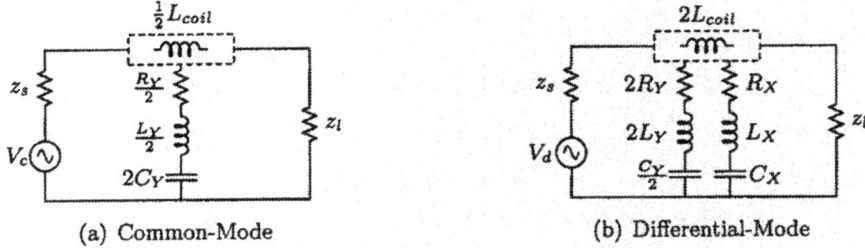

(a) Common-Mode   (b) Differential-Mode

Figure 6.20: Common- and differential-mode equivalent circuits used to simulate the filters in Section 6.3.

For simulation, both the line-to-ground (Y) capacitor (Panasonic ECK-ATS472ME6) and the line-to-line (X) capacitor (Rubycon 250MMCA334KUV) are modeled with first-order equivalent series resistance, equivalent series inductance, and bulk capacitance. The nominal value of capacitance, along with the measured values of inductance and resistance, are used in the model of each capacitor. The parameters are: $R_Y = 200 \text{m}\Omega$, $L_Y = 48.6 \text{nH}$, $C_Y = 4700 \text{pF}$, $R_X = 50 \text{m}\Omega$, $L_X = 49.4 \text{nH}$, $C_X = 330 \text{nF}$. The source and load impedances, z_s and z_l respectively, match those of the network analyzer, 50Ω. The coil used in the test filters, shown in Fig. 6.7, is represented by the inductance matrix L_{coil}, which is obtained using the numerical inductance calculation tool FastHenry:

$$L_{coil} = \begin{bmatrix} 36.52 & 42.62 & 0.236 \\ 42.62 & 224.6 & 8.089 \\ 0.236 & 8.089 & 5.262 \end{bmatrix} \text{nH}$$

For common-mode, T_3 is set to $1\text{M}\Omega$ to approximate an open circuit, and T_2 is set to the effective impedance of the Y capacitor, $Z_{Y_{CM}} = \frac{1}{2}\left(R_Y + (j\omega C_Y)^{-1} + j\omega L_Y\right)$. For the differential-mode simulation, the effective impedance of the Y capacitor is different. With the two capacitors in series, the effective impedance, and thus T_2, now becomes $Z_{Y_{DM}} = 2\left(R_Y + (j\omega C_Y)^{-1} + j\omega L_Y\right)$. The effective impedance of the X capacitor is $Z_{X_{DM}} = R_X + (j\omega C_X)^{-1} + j\omega L_X$, the value used for T_3.

The results of the common- and differential-mode simulations are shown in Fig. 6.21. Comparing the simulation results to the experimental data in Fig. 6.12, the differential-mode results do match in an absolute sense. The *Optimal Y (DM)* simulation is roughly between the two measured coupling cases, which is understandable given coupling between the coils is not modeled. However, the addition of the X capacitor in the *Optimal Y, Uncancelled X (DM)* fails to match the same downward-shift in resonance to near 20MHz, which exists in both experimental measurements. The shift of this resonance is representative of an increase in effective inductance in the Y capacitor branch, which may be a result of unmodeled inductive coupling between the X and Y capacitors. The important similarity between the experimental measurements and the simulation is seen comparing *Optimal Y, Uncancelled X (DM)* and the final trace *Optimal Y, Optimal X (DM)*. By

90

appropriately locating the X capacitor on the coil, it is possible to both shift the resonance higher in frequency, and to increase the attainable attenuation.

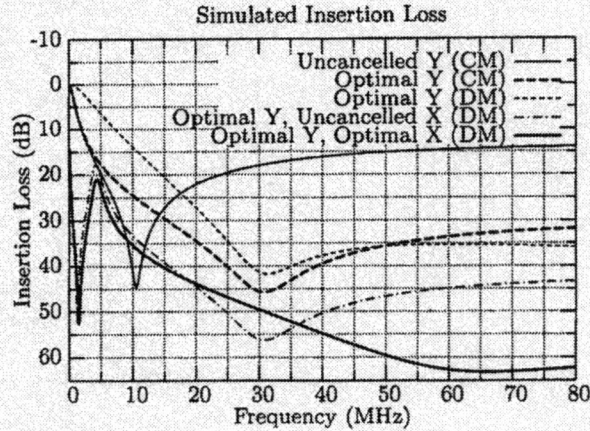

Figure 6.21: Simulated results for the filters in Section 6.3, using the circuits shown in Fig. 6.20. Note the different frequency range than in Fig. 6.12.

If consideration is given to modeling the increase in effective inductance in the Y capacitor branch when the X capacitor is present, significantly improved correlations between the experimental measurements and the model simulation result. If $Z_{Y_{DM}}$ is increased by a modest 20% to make $Z_{Y_{DM}} = 2\left(R_Y + (j\omega C_Y)^{-1} + j\omega\left(L_Y + 0.2L_Y\right)\right)$, and the tuning location of the X capacitor on the coil is slightly moved (by 1.5mm), a refined inductance matrix yields,

$$L_{coil} = \begin{bmatrix} 36.52 & 42.29 & 0.563 \\ 42.29 & 222.4 & 9.091 \\ 0.563 & 9.091 & 5.432 \end{bmatrix} nH$$

the plot in Fig 6.22 results. These results correlate significantly better than the case without the added effective inductance, although differences are still notable for *Optimal Y, Uncancelled X (DM)*.

Given the substantial modeling simplifications used in creating these simulations (e.g. neglecting coil-to-coil and other mutual couplings, using simple numerical simulations to obtain coil inductances, etc.) the degree of accuracy of the model is striking, confirming its usefulness for understanding the behavior of such systems.

6.6   Conclusion

The size and performance of discrete EMI filters are often limited by their component parasitics, such as the equivalent series inductance of capacitors. Implementing inductance cancellation traditionally requires at least one winding for each capacitor, increasing the volume and cost of the filter if all capacitor inductances are to be cancelled in a balanced fashion.

This chapter has extended the inductance cancellation presented in Chapter 1 by developing a method that allows for the use of a single magnetic winding to compensate for the effects of equivalent series inductances of two capacitors, instead of just one.

91

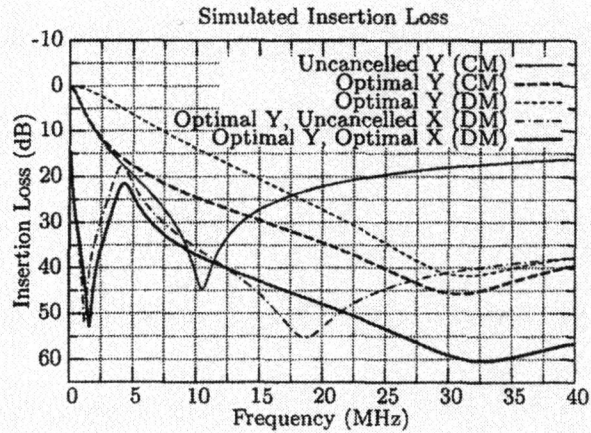

Figure 6.22: Simulated results for the filters in Section 6.3, using the circuits shown in Fig. 6.20 with additional differential-mode Y capacitor inductance.

This multiple-element compensation method was applied experimentally to both test filters and to a commercially-available EMI filter with great success. Further, the coupling of closely oriented magnetic windings was also investigated, illustrating their successful use in constrained spaces, and a possible avenue for optimizing winding size. Finally, an analytical basis for the inductance compensation is developed and compared to experimental results.

Chapter 7

Other Applications of Inductance Cancellation

7.1 Introduction

Inductance cancellation can be used in any application in which it is desirable to shift inductance from one branch to two other adjoining branches. This technique will never lower the total amount of inductance in the circuit but in some applications, such as filtering, the repositioning of inductance can enhance the performance. This tradeoff can be analyzed by examining the T-model of the transformer used in inductance cancellation. If one branch is negative, than the other two must be positive and the net inductance measured across any two terminals must be positive.

This chapter considers the use of inductance cancellation in applications other than filtering. One such application is to compensate for the parasitic inductance associated with a sense resistor. This inductance creates a zero (which is usually unwanted) within the control loop of a converter. Possible application in power electroncs circuits will also be introduced. It is expected that many other applications of inductance cancellation exist in the area of power electronics and elsewhere. This chapter also serves to point toward the possibilities in this direction.

7.2 Inductance Cancellation of a Sense Resistor

A sense resistor is a precision resistor that is used to measure the current passing through it. Measurement of the voltage across the resistor can be used to determine the current flowing in the resistor. Resistors capable of measuring high currents usually have very low values and high power ratings. A differential amplifier is usually used to determine the current though the sense resistor. A typical application of current sensing is shown in Fig. 7.1.

The transfer function of this circuit is

$$\frac{V_o}{I_{\text{circuit}}} = \frac{R_f}{R_i}(R_{\text{Sense}} + L_{ESL}s). \tag{7.1}$$

Thus the current sensor will have a zero at the frequency $\frac{R}{L2\pi}$. For example a $5m\Omega$ sense resistor (Ohmite 15FR005) is measured with an impedance analyzer across a range of frequencies. At 10 kHz a 55 nH series inductance is measured. This sense resistor would then have a zero at 14 kHz.

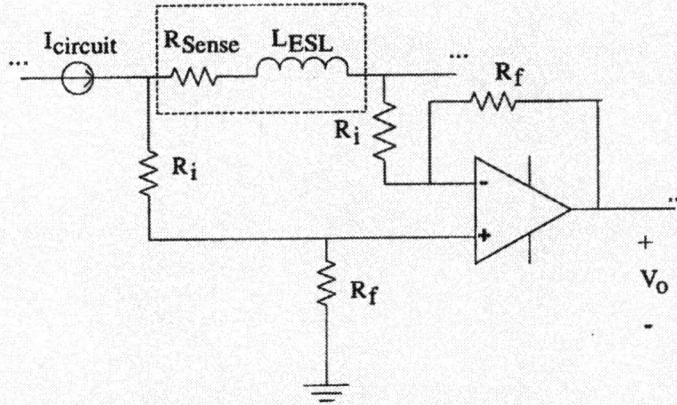

Figure 7.1: A typical current sensing scheme. A differential amplifier measures the voltage drop of the sense resistor.

Inasmuch as power converter switching frequencies (and desired sense bandwidth) are often much higher than this, some means of compensating for the parasitic inductance is desirable.

One way that the effects of this zero can be offset is by adding two matched capacitors, C, to the differential amplifier in parallel with each R_f resistors, as illustrated in Fig 7.2. These two capacitors will change the transfer function of the system to

$$\frac{V_o}{I_{circuit}} = \frac{R_f R_{Sense}}{R_i} \frac{\frac{L_{ESL}}{R_{Sense}} s + 1}{CR_f s + 1}. \tag{7.2}$$

Thus the current sensor can add a pole to the circuit to cancel out the zero, assuming CR_f equals $\frac{L_{ESL}}{R_{Sense}}$. One limitation of this approach is the inherent sensitivity of this pole-zero cancellation to component variation, (especially of the capacitors).

Inductance cancellation is another means of compensation. One of the state variables in a power converter is the inductor current. Many control schemes will have a sensor measuring (directly or indirectly) the inductor current. If the current sense resistor is in series with the inductor then inductance cancellation can easily be applied with no effect on the circuit[1]nductance cancellation may be applied even when the sense resistor is not in series with another inductance, as long as circuit operation is not greatly impacted by the introduction of additional inductance on the order of the ESL.. In the approach proposed here and illustrated in Fig. 7.3(a), the effect of the ESL of the sense resistor will be cancelled in the system output. The positive inductances caused by the inductance cancellation transformer will be in series with the load resistor in the power converter and in series with one of the input resistors to the differential amplifier. Inasmuch as the input impedance of the differential amplifier is high compared to that of the inductance that is introduced in the amplifier path (L_B) there will only be a small (and negligible) voltage ripple across the inductance.

A test setup demonstrating this approach for the sense resistor was created and is shown in Fig. 7.3b. In this circuit a power amplifier produces a sinusoidal output across a 50Ω load and a $5m\Omega$ sense resistor. The sense resistor is connected to ground, therefore the positive terminal of

---

[1]I

94

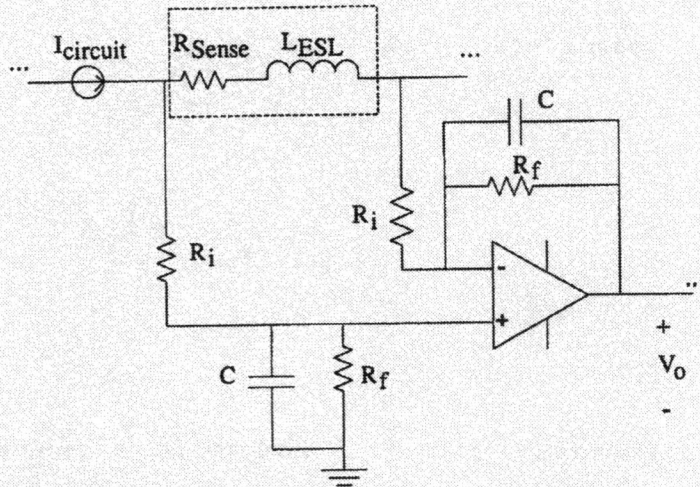

Figure 7.2: A current sense resistor and amplifier that used two capacitors to add a pole that will offset the zero caused by the inductance of the sense resistor.

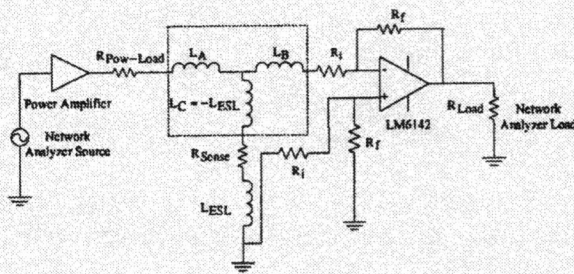

Figure 7.3: A schematic (a) and picture (b) of the test setup of a resistor with inductance cancellation.

the operational amplifier (LM6142) can also be connected to ground; a single-ended amplifier thus suffices in this application. The amplifier has a gain of 1 so that it can be considered ideal for the frequency range of interest.

In this configuration the current sensor has a low frequency zero due to the inductance associated with the $5m\Omega$ sense resistor. The sensed signal due to a constant magnitude sweep of input frequencies shows that the sensed signal increases as the input frequency increases. A pole is introduced into the system by adding a $2.2nH$ capacitor in parallel with the feedback resistor, as illustrated in Fig. 7.2. The pole cancels out the zero that exists in the system and the sensed signal is constant across the frequency range. The ratio of the current and the output voltage from the operation amplifier is, as expected, $.005\Omega$. A comparison of the sensor performance across frequency with and without capacitor compensation of the ESL is shown in Fig. 7.4. Note that the system without the compensation capacitor exhibits the effect of the unwanted zero for the entire frequency range. The minimum frequency for the power amp is 10 kHz and the zero for this system

95

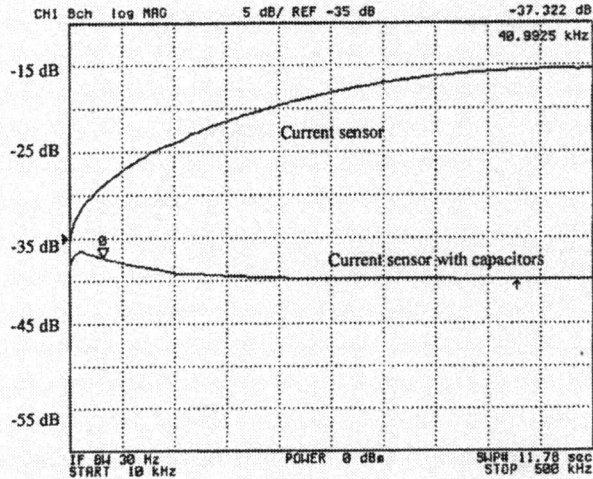

Figure 7.4: The sensed output voltage due to a constant magnitude input frequency sweep of a current sense resistor and an operational amplifier with and without a pole at low frequencies.

is at approximately this frequency.

Inductance cancellation was also used to cancel the inductance of the sense resistor. The sense resistor is wound with a center-tap transformer configuration. The transformer is wound about the sense resistor in a fashion similar to the capacitors described in Section 4.3. The backed copper foil is 525 mils wide and 10 mils thick with 1 mil of insulation. The transformer has 3 turns on the first set of windings and 1.25 turns on the second set of windings. The positive inductances that are introduced by the inductance cancellation transformer are either in series with the 50Ω power load are in series with R_i. The negative inductance is in series with the sense resistor and cancels out the effect of the ESL in the sense path. The measured output signal is constant in magnitude for a range of input frequencies as shown in Fig. 7.5. Note that ratio of the current and the output voltage from the operation amplifier is higher than in the previous case. The added resistance of the inductance cancellation transformer is comparable to the sense resistor. In a manufactured part this resistance would be incorporated into the desired value of the sense resistor. Alternatively, a convenient sense resistor could be employed with a judicious circuit trace layout that provides the correct magnetic coupling for cancellation much as is done for filter capacitors in Chapter 3.

Two methods were described to compensate for the parasitic inductance of the sense resistor. Both methods work and are inexpensive to use: adding two small value capacitors or using inductance cancellation. One difference between the two approaches is that adding capacitors to the differential amplifier only places a pole nearby the zero, and matching the pole and zero frequency will be a function on the tolerances of all the parts. With imperfect matching there may be a long tail transient. When a resistor with inductance cancellation is used the zero is moved to a higher frequency depending on how much in the inductance is cancelled.

96

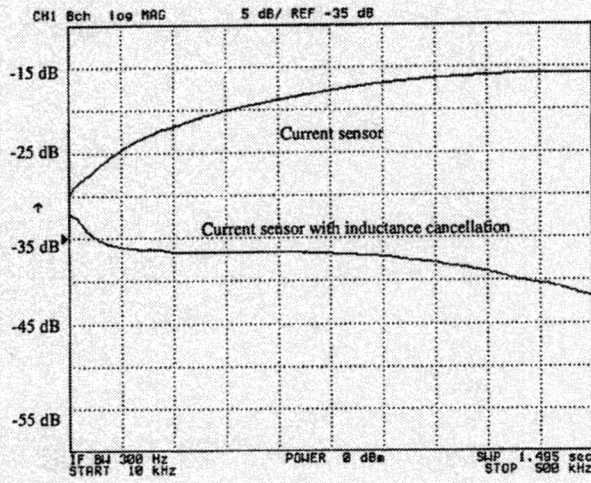

Figure 7.5: The sensed output voltage due to a constant magnitude input frequency sweep of a current sense resistor with and without inductance cancellation and an operational amplifier.

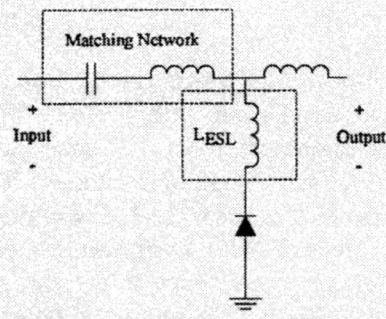

Figure 7.6: A rectifier circuit in which the equivalent series inductance of the diode can be adjusted with inductance cancellation.

7.3   Inductance Cancellation In Power Electronic Circuits.

In some applications its possible to use inductance cancellation to compensate for the parasitic inductance associated with the leads of a switching device. In some resonant and high frequency power conversion topologies, for example switching devices are placed such that inductance cancellation can be gainfully applied. This section will offer three examples of such circuits.

The first example is high frequency rectifier which has a diode in the shunt path and is connected to two inductors as shown in Fig. 7.6. The sinusoidal input which drives the system incorporates a resonant or matching network. The rectifier is made up of the diode, and an inductive output ensures a constant current to the load. Inductance cancellation can be used to adjust or cancel the lead inductance of the diode. The other positive inductances will be in series with the other two inductors in the circuit. In this application, it can be shown that judicious adjustments of the shunt path inductance (including device package inductance) can improve converter performance.

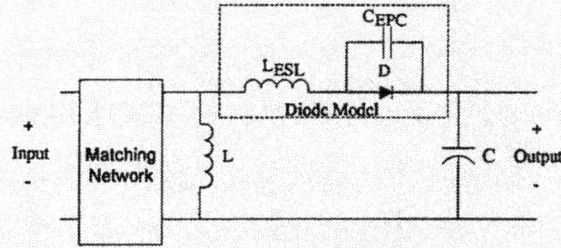

Figure 7.7: A second rectifier circuit in which the equivalent series inductance of the diode can be adjusted with inductance cancellation.

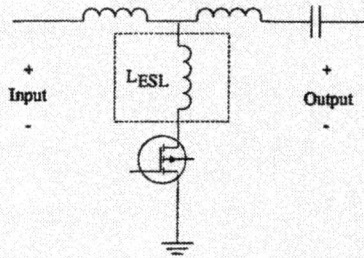

Figure 7.8: The class E amplifier can also use inductance cancellation to adjust the equivalent series inductance of the MOSFET.

The second example is another rectifier. In this case the diode is in the series path and one of the inductors is in the shunt path, as shown in Fig. 7.7. The input of this rectifier is also a resonant or matching network. If inductance cancellation is applied to this circuit the equivalent series inductance of the diode can be adjusted or cancelled (its possible that a lower, but non-zero inductance will improve operation of the inverter although this has not been confirmed at the time of this writing). The other positive inductances introduced by inductance cancellation will be in series with the shunt path inductance and the matching network and can form part of all of these networks.

The last example is the class E amplifier shown in Fig. 7.8. The input of the amplifier will be a dc voltage and the output will be an ac waveform. The duty cycle of the MOSFET is typically 0.5 and the switching frequency is usually very high. In the class E amplifier inductance cancellation can be used to adjust the effective inductance of the MOSFET; and the additional positive inductances of the T-model can be absorbed as part or all of the other two branch inductances in the circuit. Similar approaches can be used for related topologies, such as class F amplifiers.

In all three examples shown, there exists a node with three branches containing inductances. One of these branches consists of a parasitic inductance associated with a switching device whereas the other two branches have discrete inductors. By using inductance cancellation the effective parasitic inductance in series with the device can be adjusted to values below that of the package. The additional (positive) inductances associated with the cancellation transformer are absorbed as part (or all of) the other branch inductances used in the circuit.

7.4 Conclusions

Inductance cancellation has useful applications beyond filter design. Any time a three branch node could benefit from shifting the inductance from one branch to the other two branches, then inductance cancellation can be used.

This chapter showed an example of a sense resistor circuit is which inductance cancellation is used to eliminate an unwanted zero. The zero caused by the equivalent series inductance of the sense resistor can be alternatively compensated by two appropriately chosen capacitors for pole-zero cancellation. Inductance cancellation is useful since it effectively *moves* the zero in this system to a very high frequency rather than relying on a pole-zero cancellation. In special cases of power electronics applications inductance cancellation can be used to move some or all of the inductance associated with a switching device to adjacent branches.

Chapter 8

Parasitic Capacitance Cancellation in Filters

8.1 Introduction

Filter inductors and common-mode chokes suffer from both parasitic resistance and capacitance. Winding resistance and core loss lead to parasitic resistance, while parasitic capacitance arises from capacitance between winding turns and from winding-to-core capacitance. The distributed parasitic components can be lumped together to form the lumped-parameter model for an inductor shown in Fig. 8.1a [48–51]. The impedance magnitude of a practical inductor as a function of frequency is illustrated in Fig. 8.1b. The parasitic capacitance dominates the impedance above the self-resonant frequency of the inductor (typically 1-10 MHz for power electronics applications, but sometimes lower for ungapped common-mode chokes). This parasitic capacitance reduces the impedance of an inductor at high frequencies, and hence reduces its effectiveness for high frequency filtering.

   This chapter introduces a technique for improving the high-frequency performance of filter inductors by cancelling out the effects of the parasitic capacitance. This technique uses additional passive components to inject a compensation current that cancels the current flowing through the parasitic capacitance, thereby improving high-frequency filtering performance. The proposed technique is related to strategies that have been exploited for reducing common-mode noise in certain power supply topologies [52–57], and is applicable to a wide range filtering and power conversion designs where the parasitic feedthrough of magnetic components is an important consideration.

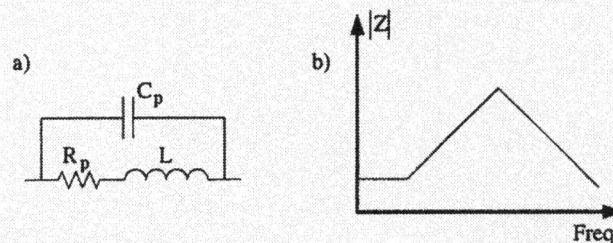

Figure 8.1: A simple inductor model including parasitic effects. An impedance vs. frequency plot shows that the capacitance limits the impedance at high frequencies.

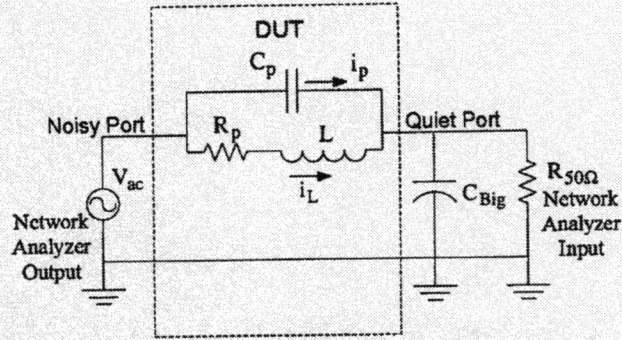

Figure 8.2: A test circuit for evaluating the filtering performance of magnetic components. The device under test (DUT) is a filter inductor.

8.2 Capacitance Cancellation

The proposed technique improves the performance of magnetic components (e.g., inductors and common-mode chokes) in filter applications where the function of the component is to prevent the transmission of high-frequency current from a "noisy" port to a "quiet" port. We assume that the "quiet" port is shunted by a sufficiently low impedance (e.g., a capacitor) that it is effectively at ac ground, and that small amounts of high-frequency current into the "noisy" port are acceptable so long as they are not transmitted to the "quiet" port. These assumptions are satisfied in a wide range of filtering and power conversion applications. A test circuit for evaluating the attenuation performance of filters is illustrated in Fig. 8.2. The "noisy" port of the device under test is driven from the output of a network analyzer, and the response at the "quiet" port is measured at the 50 Ω input of the network analyzer.

The capacitance cancellation technique developed here is not geared towards changing the parasitic capacitance itself (i.e., the capacitance C_p of the device under test in Fig. 8.2). Rather, the deleterious effect of the parasitic capacitance - the current that passes through it at high frequencies - is nullified by a counterbalancing current injected at the quiet port. This counterbalancing current is injected by passive circuitry introduced expressly for this purpose. Referring to Fig. 8.2 (where an inductor and its parasitics form the device under test), the quiet port is assumed to be at ac ground, so the voltage across the device under test is the ac component of the voltage at the noisy port. The capacitor current, i_p, injected into the quiet port by the noisy port voltage v_{ac} at the angular frequency ω can thus be expressed as:

$$i_p = j\omega v_{ac} C_p \tag{8.1}$$

The goal of cancelling the current i_p at the quiet port can be achieved by adding an additional compensation winding and capacitor to the inductor, as illustrated in Fig. 8.3a. The compensation winding carries only small, high-frequency currents, and can be implemented with very small wire. As shown in Fig. 8.3b, the compensation winding forms a transformer with the main winding, with the magnetizing inductance operating as the filter inductor. Neglecting other transformer parasitics (such as leakage) the circuit operation can be described as follows: The net current injected into the quiet port can be calculated as:

101

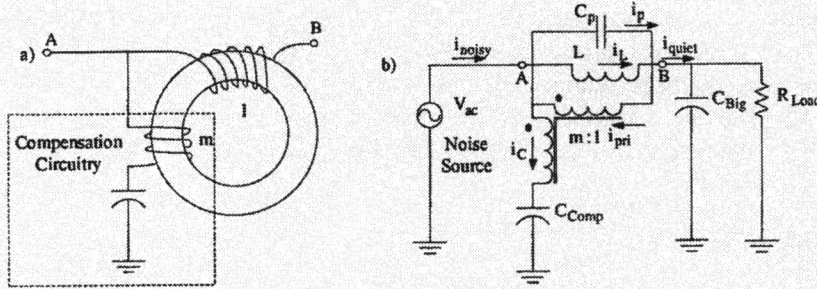

Figure 8.3: (a) An additional winding and a capacitor are added to the inductor to form the circuit for parasitic capacitance cancellation. (b) equivalent circuit of (a) in the test circuit of Fig. 8.2.

$$i_{\text{quiet}} = \frac{v_{ac}}{j\omega L} + j\omega v_{ac}[C_p - m(1-m)C_{\text{Comp}}] \tag{8.2}$$

where m is the ratio of the compensating winding turns to the main winding turns, and C_{Comp} is the value of the compensation capacitor. By selecting the winding ratio and compensation capacitor such that $m(1-m)C_{\text{Comp}} = C_p$, the parasitic capacitance current injected into the quiet port is cancelled by the compensation circuit, leaving only the inductive current component.

While this technique eliminates the effect of the parasitic capacitance at the quiet port, it does not eliminate the parasitic effects in other regards. Most notably, the current into the noisy port actually increases at high frequencies. For any selection of winding ratio and compensation capacitor described above, the currents i_p and i_{pri} will ideally cancel out and the current from the noise source will be $i_L + i_C$ or

$$i_{\text{noisy}} = \frac{v_{ac}}{j\omega L} + j\omega v_{ac}C_p\frac{1}{m} \tag{8.3}$$

Since m is selected to be < 1 to achieve cancellation at the quiet port, the capacitive current into the noisy port increases somewhat as compared to the uncancelled case. The additional current is the same as that of a small capacitor bypassing the noisy port. In many applications this is quite acceptable, and it becomes advantageous in filter applications where a capacitive bypass of the noisy port is desirable.

The proposed compensation technique is related to a number of other filtering and balancing techniques that have been explored in the past. The *topology* of the filter network that is created (e.g., Fig. 8.3b) is identical to coupled inductor filter structures that have been used widely for everything from notch-filtering [4, 6] to "zero ripple" filters [5, 7–10] to filters incorporating parasitic inductance cancellation [58, 59]. However, the design goals and component values of these other circuit implementations are vastly different from the case considered here. The technique presented here bears a closer relationship to recent work on cancellation of common-mode noise in certain switching-power supply topologies [52–57]. These use additional magnetics and capacitors to compensate common-mode currents injected from switching devices through the capacitance to the circuit ground or enclosure. The approach proposed here is different in that it focuses on parasitic capacitance appearing across magnetic components, and applies to a broad range of topologies and applications. Interestingly, these cancellation techniques are also related to the "neutralized" amplifier configuration developed more than fifty years ago, in which injected currents were used to compensate Miller capacitance effects in vacuum tube audio amplifiers [60, 61]. In this chapter, we focus on the use of capacitance cancellation to enhance the performance of electrical filters. It is expected that the proposed approach will be less costly and burdensome to implement than more conventional EMI shielding techniques (e.g., a faraday shield) in many applications.

8.3 Evaluation

To validate the proposed approach, a number of prototype circuits have been constructed and tested. For simplicity, the prototype circuits we describe here employ $m = 0.5$ and $C_{\text{Comp}} = 4C_p$ except as otherwise specified. This turns ratio can be easily achieved with a 1:1 center-tapped transformer by using the full winding of the primary and only half the winding on the secondary. The effects of varying the turns ratio will also be examined.

A prototype filter circuit with capacitance cancellation was constructed using one half of a P3219-A common-mode choke (Coilcraft, Inc., Cary, IL). Only one winding of the choke was used in the testing; the other winding was left unconnected. The main winding of this toroidal choke comprises 45 turns of 25 gauge wire in a single layer, forming a 10.8 mH filter inductance, with approximately 16 pF of parasitic capacitance. A compensation winding of 22 turns of AWG 30 wire was added to the choke, as illustrated in Fig. 8.3a. A range of compensation capacitor values were tested to identify the best value and an 18 pF ceramic capacitor was selected. A 3.3 nF ceramic capacitor was used to bypass the quiet port.

The filter circuit was evaluated using the network analyzer-based test setup illustrated in Fig. 8.2. The filter circuit is driven from the network analyzer output, and the response at the quiet port is measured at the 50 Ω input of the network analyzer. Figure 8.4 shows the magnitude of the transfer function from the noisy port to the quiet port over the conducted EMI frequency range (up to 30 MHz) both with and without parasitic capacitance cancellation. As can be seen, the proposed cancellation technique improves performance by approximately 7 dB in the frequency range of 10 MHz to 28 MHz; tests on other devices provided similar results - a 5-10 dB improvement in performance at high frequencies. Most of the irregular peaks in the figure are due to unmodelled, higher frequency parasitics in the filter.

While the proposed cancellation technique does improve filter performance, the results are not as good as might be expected. The main source of this limitation is the leakage inductance of the cancellation transformer. All impedances in series with the compensation capacitor, especially the leakage inductance, will compromise the high frequency performance of the cancellation. Treating the leakage inductance of the compensation winding as part of a compensation impedance Z_C and ignoring the primary side leakage inductance, the quiet port current becomes

$$i_{\text{quiet}} = \frac{v_{ac}}{j\omega L} + j\omega v_{ac} C_p - \frac{m(1-m)}{Z_{\text{Comp}}(j\omega)} \tag{8.4}$$

$$i_{\text{quiet}} = \frac{v_{ac}}{j\omega L} + v_{ac} \frac{C_p C_C L_C (j\omega)^3}{C_C L_C (j\omega)^2 + 1} \tag{8.5}$$

where L_C is the secondary leakage inductance in series with C_C. The term $C_C L_C (j\omega)^2$ greatly affects the performance of capacitance cancellation. C_C is minimized when the turns ratio, m, is 1/2 and L_C is dependant on the magnetic element and is lower for transformers with a lower turns

103

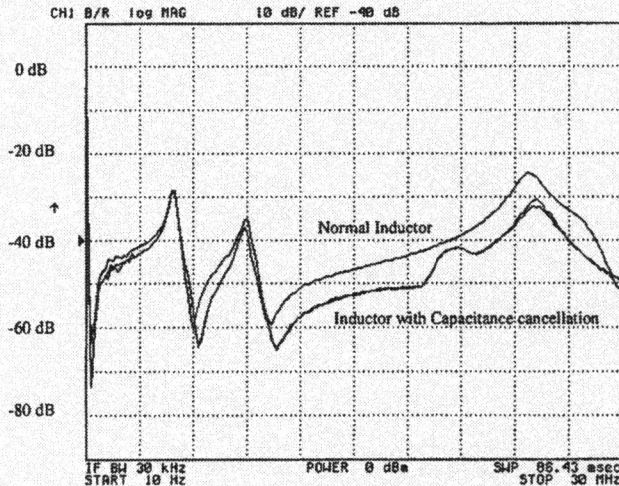

Figure 8.4: Ratio of input to output voltage for the P3219-A Coilcraft choke. The higher curve is the response without the capacitance cancellation. Above 10 MHz there is about 7 dB of improvement from using the capacitance reduction technique.

ratio. Thus the coupling coefficient of the primary and cancellation windings is a critical factor in achieving good cancellation performance.

The transformer used in Fig. 8.4 was rewound such that the secondary has three turns resulting in a turns ratio of 0.06. The compensation capacitor was chosen to maximize the impedance of the inductor at frequencies lower than 6 MHz; the best compensation capacitors were experimentally determined to be 56 pF and 59 pF when the turns ratios are 0.5 and 0.06. Note that 56 pF is much larger than the expected compensating capacitor for a turns ratio of 0.5. However, this capacitance, when considered with the leakage inductance in the compensation branch, leads to the system with the best performance at the frequencies of interest. Assuming that the transformer with 10.8 mH primary inductance has a coupling coefficient is 0.98 in both cases, the value of $C_C L_C$ changes from $3.024 \cdot 10^{-15}$ H F to $3.81 \cdot 10^{-17}$ H F when the turns ratio changes from .5 to .06. A comparison of the performance of the two cases of capacitance cancellation is shown in Fig. 8.5. This figure shows that the system with the lower turns ratio outperforms the other system at frequencies below 6 MHz, and using a lower turns ratio will raise the self-resonant frequency from 400 kHz to 1.4 MHz. The compensation capacitors were chosen in both cases to improve performance below 6 MHz. Figure a) shows this frequency range. Signal 1 is the result of the normal inductor and signal 2 is the result of the inductor with capacitance cancellation with a turns ratio of 0.5 and a compensation capacitor of 56 pF. The system is changed such that the turns ratio is 0.06 and the compensation capacitor is 59 pF and the performance is improved to signal 3. Figure b) shows that above 6 MHz the performances of all three systems are dominated by other parasitics and that each system will outperform the others over some frequency range.

The model of the inductor used in Fig. 8.5 cannot be accurately described using Fig. 8.1 for the frequency range of interest. All three positive sloped regions in Fig. 8.5 (300 kHz - 5MHz, 6 MHz - 8.5 MHz, and 10 MHz - 22 MHz) represent areas in which the inductor looks capacitive, each region with a different capacitance. (A more accurate model of the inductor includes an two

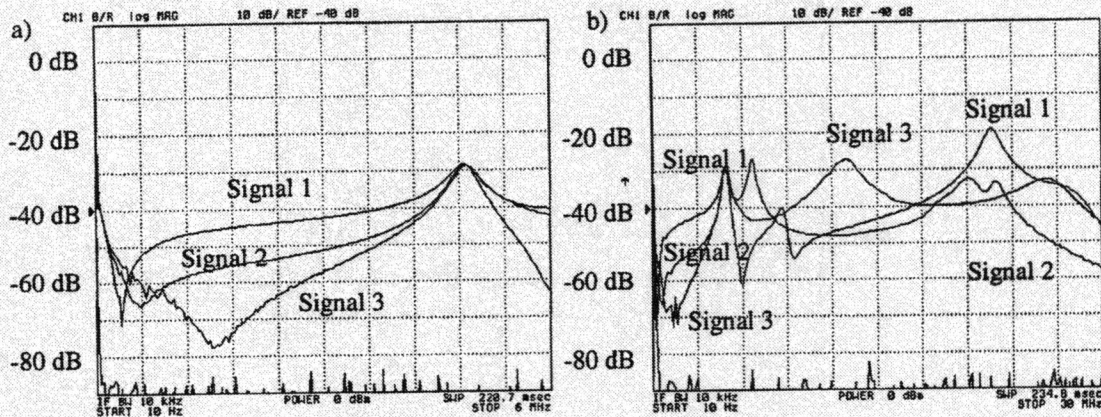

Figure 8.5: Ratio of output to input voltage for the P3219-A Coilcraft choke in which the turns ratio of the transformer is changed from 0.5 to 0.06.

LC branches in parallel with the existing model of Fig. 8.1.) Capacitance cancellation can target any one of these three capacitances but with different values of the compensation capacitor, but during the other regions the capacitance cancellation winding will either not be optimal, or in the case of Fig. 8.5, harmful for the operation of the filter. Thus, depending on the frequency range of interest the compensation capacitor can be chosen.

In order to quantitatively evaluate the effects of the coupling coefficient on performance, a PSpice model of the experimental system leading to Fig. 8.4 was created (without the higher order parasitics). The parasitic capacitance of the inductor was set to 15 pF and the coupling coefficient was varied between .98 and 1. The simulated filter attenuation at 20 MHz is plotted as a function of the coupling coefficient in Fig. 8.6. This figure shows that the effectiveness of capacitance cancellation is very sensitive to the coupling coefficient. These insights are also relevant to other passive EMI cancellation techniques, such as those in [52–55].

This test was expanded to take into consideration changes in the turns ratio and changes in the amount of inductance and capacitance of the inductor under test. Fig. 8.7 shows a normalized plot of the filter attenuation at 20 MHz as a function of the term $C_{Comp}L_{Comp}$. Each point on the plot shows a set of simulations with different leakage inductances with a unique turns ratio, magnetizing inductance, and parasitic capacitance. Since some of the systems have different parasitic capacitances the total amount of possible attenuation improvement will be different. Thus for each set of simulations the amount of attenuation is given as a percentage of the maximum possible case (15 to 30 dB depending on the case). The three outlying cases marked with diamonds represent cases in which the turns ratio is 0.125, otherwise a clear relationship exists between the relative improvement in attenuation and the term $C_C L_C$. Further analysis of the data used to generate this plot shows that as the turns ratio, m, decreases, the range of the transformer's coupling coefficients over which beneficial results occur will increase. Hence, transformers with lower coupling coefficients benefit from a low turns ratio. The higher lines in the figure, though, correspond to systems with high turns ratios and high coupling coefficients, thus a transformer with near ideal coupling will lead to the overall best system possible.

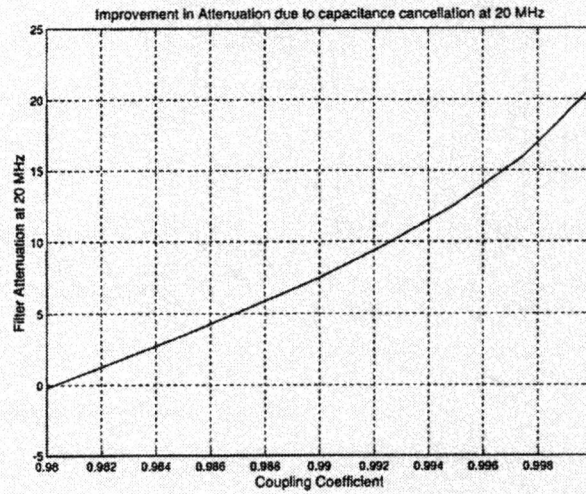

Figure 8.6: The effects of changing the coupling coefficient on the system in Fig. 8.2. The parasitic capacitance of the inductor is 15 pF, the magnetizing inductance is 50 μH, and the turns ratio (m) is 0.5.

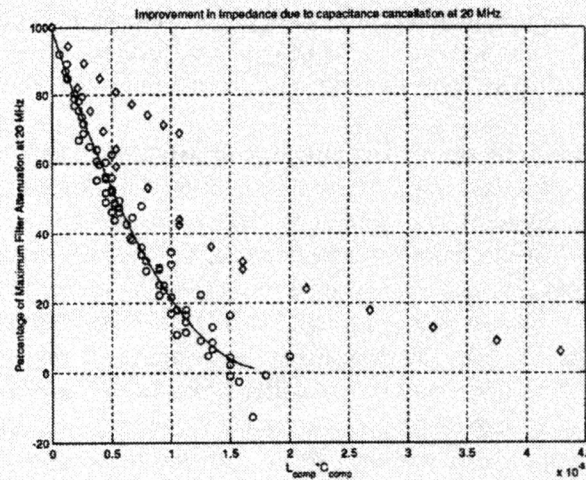

Figure 8.7: The effects of changing the term $C_{\text{Comp}}L_{\text{Comp}}$, the total series capacitance and inductance in the compensation path, on the improvement of attenuation at 20 MHz. Each point corresponds to a specific magnetizing inductance, parasitic capacitance and turns ratio of the transformer.

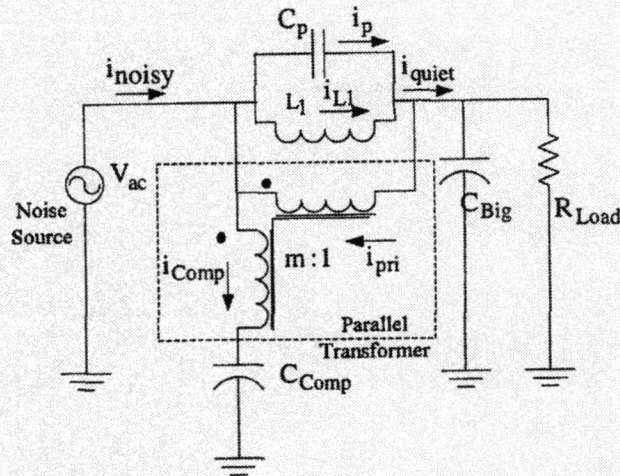

Figure 8.8: An equivalent circuit model showing how capacitance cancellation can be implemented using a parallel RF transformer to inject cancellation currents.

8.4 Alternative Implementation

Here we introduce an alternative implementation of the capacitance cancellation technique that avoids the magnetic coupling limitations of the simple implementation. Rather than relying on the limited magnetic coupling achievable with an additional winding on the filter inductor, the alternative implementation achieves the cancellation using a separate radio-frequency (RF) transformer in parallel with the filter inductor, as illustrated in Fig. 8.8. The compensation capacitor is selected to compensate the parasitic capacitances of both the transformer and inductor. As will be shown, this technique can achieve highly effective capacitance cancellation using only small, inexpensive components.

The tradeoffs that arise in designing the inductor/transformer combination are similar to those that occur in some types of active EMI filters [62]. These design tradeoffs are summarized here. First, the RF transformer should have a high coupling coefficient to enable good capacitance cancellation to be achieved. The transformer magnetizing inductance should be similar to or larger than the original filter inductance to realize the desired filtering performance because it is desirable to have the majority of the current going through the original filter. Since it only needs to carry the small cancellation currents, the RF transformer can be wound with very fine wire, and can be made quite small and inexpensive. Differences in winding resistances between the RF transformer and filter inductor can be used to ensure that low-frequency currents flow through the filter inductor, though a small blocking capacitor could also be introduced for this purpose. One RF transformer meeting these criteria is the WB3010-1 (Coilcraft, Cary, IL). This center-tapped transformer has a magnetizing inductances of 760 μH and a coupling coefficient exceeding 0.999 in a six-pin DIP package. Because the transformer is center-tapped, a 2:1 transformer is easily be created (i.e., by connecting the full winding of the primary and only half the winding on the secondary) making this design suitable for capacitance cancellation applications.

The system shown in Fig. 8.8 was tested in the test setup of Fig. 8.2. The filter inductor is one half of a 1 mH common-mode choke constructed with 12 turns fully packed on an ungapped RM12

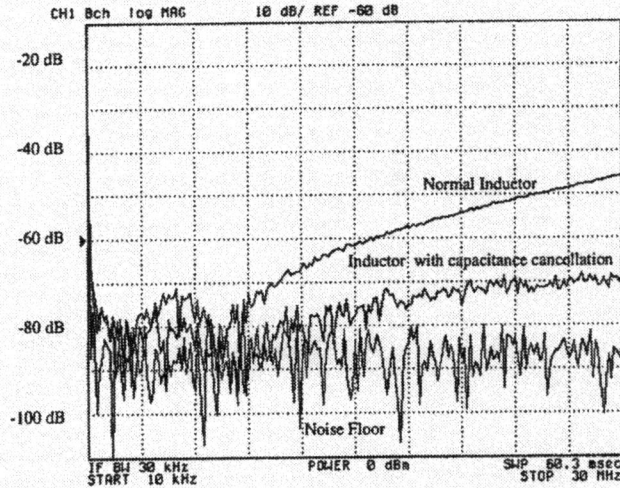

Figure 8.9: Performance of a filter inductor with and without capacitance cancellation. The inductor (L_1 as shown in Fig. 8.8) is a packed RM12 core with 1 mH of inductance and the cancellation transformer is a Coilcraft WB3010-1.

core. The 1 mH inductor, developed for automotive filtering applications, has approximately 20 pF of parasitic capacitance. The quiet-port bypass capacitance comprises the parallel combination of a 10 μF tantalum capacitor and a 1 μ F ceramic capacitor. The transformer in parallel with the inductor is the WB3010-1, and an 86 pF ceramic capacitor is used. Experimental results for this prototype system are shown in Fig. 8.9. The highest curve shows the performance without the compensation circuit. The middle trace shows the performance with the capacitance cancellation. The lowest curve shows the noise floor of the network analyzer (the measured response with the test circuit disconnected). Capacitive cancellation improves the attenuation by 25 dB at 30 MHz.

Capacitance cancellation was also applied to improve the common mode performance of an EMI filter. A commercial EMI line filter, the Schaffner FN2010-6-06, is used to test capacitance cancellation. The filter is a one stage filter and is shown in Fig. 8.10. The parasitic capacitance, as measured across either the positive or negative terminals, is 31 pF. A small center-tapped transformer, EPC3115-7 from PCA Electronics Inc., is added outside the filter case. This transformer has a turns ratio of 1:1.5, so the secondary is connected across the filter and the primary is used for the compensation path. A picture of the modified circuit is shown in Fig. 8.11. A variety of compensation capacitors were experimentally tested and a ceramic 100 pF capacitor was chosen. This capacitor is connected from the center-tap of the transformer primary to the ground (i.e., the enclosure). The modified circuit and the common mode test setup are shown in Fig. 8.10.

The resulting circuit has better performance at high frequencies. The parasitic capacitance of the filter choke is the dominant limit to attenuation at frequencies over 1 MHz until the self-resonance of the Y-capacitors (The capacitors from either line to ground) at 23 MHz. In this range of frequencies, the filter with parasitic capacitance cancellation has better performance. An improvement of at least 10 dB in attenuation is achieved from 3 to 20 MHz and an improvement of 20 dB is achieved at 14 MHz. Results of this test are shown in Fig. 8.12.

With this configuration a common mode input signal will cause a differential mode signal across

108

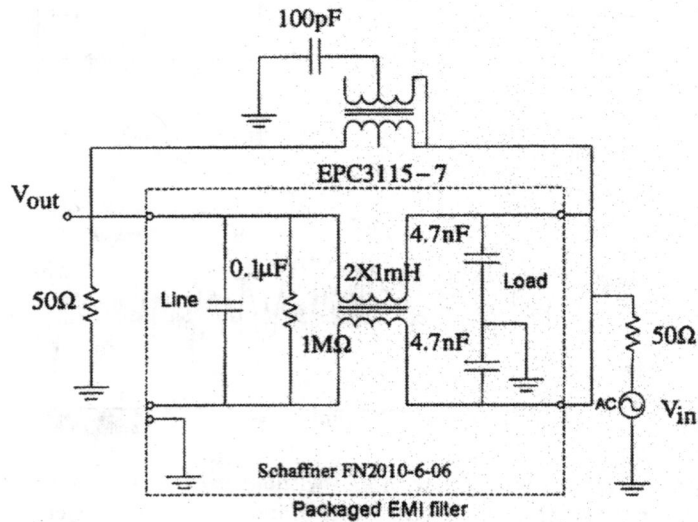

Figure 8.10: The schematic for the common mode choke with capacitance cancellation and setup for a common mode test. The two added elements, the EPC3115-7 transformer and the 100 pF capacitor, are located on the case of the filter. A common mode signal is introduced from the load side of the filter and the common mode noise is measured on the line side of the filter.

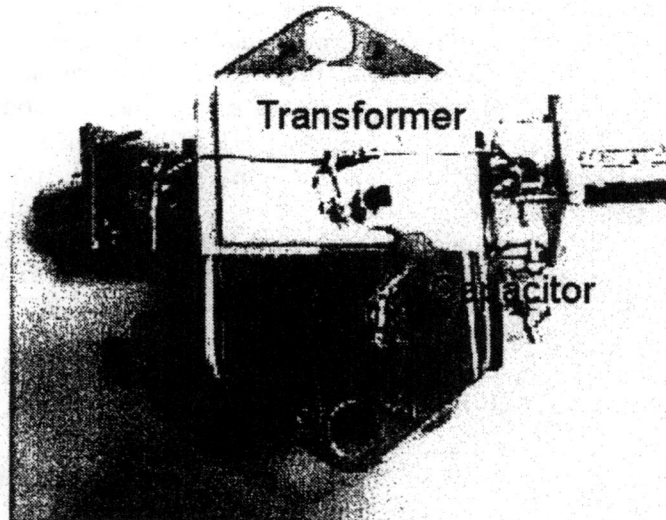

Figure 8.11: A photograph of the FN2010-6-06 filter with external capacitance cancellation. Only a small transformer and ceramic capacitor are needed to improve performance.

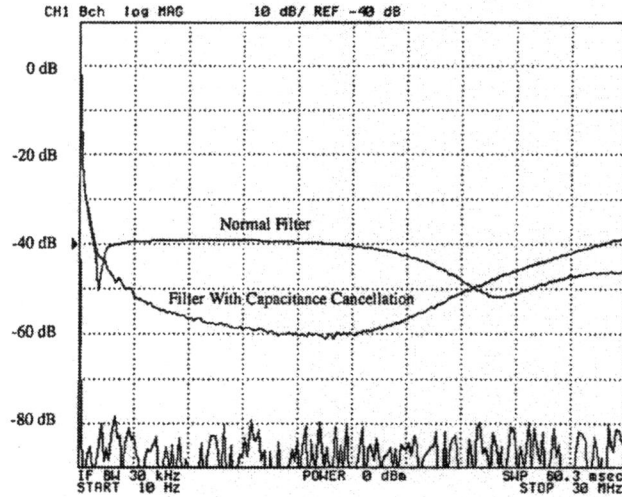

Figure 8.12: A plot comparing the performance of the common mode response of an EMI filter, Schaffner FN2010-6-06, and one modified with capacitance cancellation as shown schematically in Fig. 8.10.

the load. An ideal PSpice simulation of the system indicates up to a 64 mV differential signal may occur on the output due to a 1 V common mode disturbance as shown in Fig. 8.13. In this plot the waveform for the unmodified filter is effectively at zero, except for some spikes that are due to roundoff error (i.e., the spikes in the figure show the difference between the values PSPICE considers zero and the smallest positive number, subtracting two near identical signals will produce either of these values). The filter with capacitance cancellation on one side of the choke has significantly more differential noise. If capacitance cancellation is applied evenly to both sides of the choke then there will be no differential mode noise. When a differential mode input is applied to the circuit there is no significant change in performance due to capacitance cancellation.

The use of capacitance cancellation does have some side effects. As mentioned previously, the current from the noisy port will increase due to capacitance cancellation. Although the parasitic capacitance can be cancelled, ideal performance at all frequencies is not practical. Ideally an inductor with no parasitic capacitance will have an impedance proportional to frequency. Thus ideally, assuming a capacitor with no parasitic inductance, an L-section filter will have a 40 dB roll off from the resonant frequency onward. Real performance is not this good, mainly do to the impedance of the compensation branch. The compensation branch impedance consists of the real capacitor, leakage inductance and parasitic resistance in series. Assuming that all the other parasitic elements are negligible then the voltage $v_{ac}(1 - m)$ appears across the compensation branch. The current induced in the secondary of the transformer (and therefore proportional to the current in the primary) will be controlled in order to divert the current "traveling" in the parasitic capacitor away from the quiet port. At low frequencies the compensation impedance is dominated by the capacitance and thus the current in the primary winding of the transformer is $v_{ac}(1-m)mC_{\text{Comp}}j\omega$ which is equal to the current due to the parasitic capacitance (i.e., $v_{ac}C_p$) assuming the proper turns ratio and compensation capacitor.

At higher frequencies the impedance of the leakage inductance of the transformer will dominate the compensation impedance. When the compensation impedance is inductive then the current

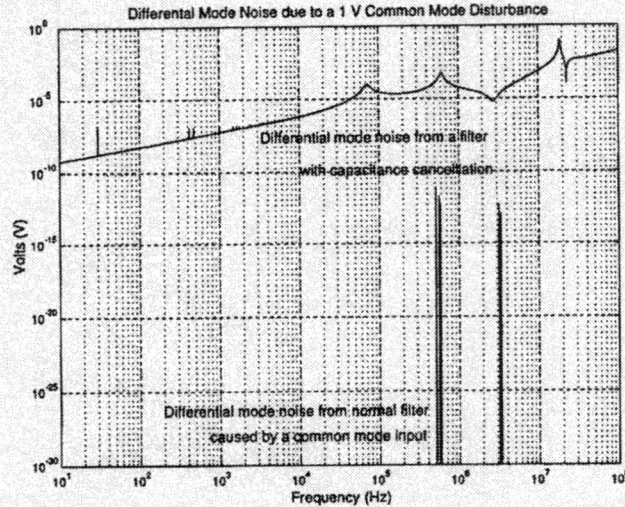

Figure 8.13: A PSpice simulation examining the differential mode noise that results from a common mode disturbance for the system shown in Fig. 8.10. The performance of the unmodified filter is essentially zero with spikes that are due to roundoff error (i.e., the spikes in the figure show the difference between the values PSPICE considers zero and the smallest positive number, subtracting two near identical signals will produce either of these values).

in the primary winding of the transformer is $v_{ac}(1 - m)m\frac{1}{L_{l_s}j\omega}$. This current will not decrease, but rather increase the amount of unwanted current going into the quiet port. Fortunately as frequency increases this current will decrease but as seen in Fig. 8.10 there will be a frequency range in which the normal inductor outperforms the inductor with capacitance cancellation. Other than the secondary leakage inductance, capacitance cancellation will have parasitic components (such as primary leakage inductance and capacitance from the primary to secondary winding of the transformer) that will also affect high frequency performance.

8.5 Conclusions

This chapter introduces a technique for improving the high-frequency performance of filter inductors and common-mode chokes by cancelling out the effects of parasitic capacitance. This technique uses additional passive components to inject a compensation current that cancels the current flowing through the inductor parasitic capacitance, thereby improving high-frequency filtering performance. The main problem with realizing capacitance cancellation is creating a compensation impedance that exactly corresponds to the parasitic capacitance. Any extra inductance in this path, which can easily be created by the leakage inductance of the transformer, will impair performance. More specifically, the performance of the system is related to the ratio of inductance reactance to capacitive reactance in the compensation branch. Two implementation approaches for this technique are introduced. The first implementation achieves cancellation using an additional small winding on the filter inductor and a small capacitor. This approach is effective where either very high coupling of the windings can be achieved or the turns ratio is very low. The second implementation uses a small RF transformer in parallel with the filter inductor to inject cancellation currents from the compensation capacitor. This technique requires an additional component (the transformer), but can provide a high degree of cancellation. Experimental results confirm the theory in both implementations and the application of this technique to a common mode EMI filter, and improvements in attenuation of up to 20 dB have been shown.

Chapter 9

Summary and Conclusions

9.1 Report Summary

Power filters are an important part of most electronic systems. The high frequency behavior of a filter is dependent on and limited by the magnitude of the parasitic components of the passive elements of the filter. These parasitic elements arise due to the physical shape and layout of the filter and can not be entirely eliminated by better design of the components. As demonstrated here, the effects of the parasitics on filter performance can be greatly reduced through parasitic "cancellation" or "compensation" techniques. This results in much higher filter performance than is otherwise achieved. This report examined two new techniques useful for filter design: inductance and capacitance cancellation.

This report contains eight chapters. The first chapter introduces the problems that arise in filter designs due to parasitic elements in the components and provides motivation for the improvement of passive elements. Chapters two through seven examine inductance cancellation, and Chapter eight discusses capacitance cancellation. Finally, this chapter concludes the report.

Chapter two provides an introduction to inductance cancellation. It describes how and why inductance cancellation works, and introduces the use of end-tapped and center-tapped transformers for this purpose. It proposes two different methods of applying inductance cancellation: embedding the cancellation magnetics on the pcb or integrating them into the element with the unwanted parasitic inductance. It proceeds though many experimental validation and example applications of inductance cancellation using a variety of different types of capacitors, both types of transformer windings, and both construction methods. It is shown that increases in attenuation of an order of magnitude or more are achievable with this approach.

Chapter three focuses on the design of air core transformers that can be layed out on the pcb and placed under a discrete capacitor. Design of such inductance cancellation transformers is explored, and applicable design rules are established and experimentally validated. The high performance of the proposed inductance cancellation technology is demonstrated in an EMI filter application.

Chapter four examines a second method of creating a component with inductance cancellation: integrating the transformer directly into the element and making a three terminal integrated filter element. This chapter investigates two different methods of constructing integrated filter elements. The second method that is proposed has a transformer that is designed in a similar manner to the pcb transformer and can be fabricated in a simple automated process. This leads to a type of integrated filter element that costs only slightly more than the capacitor that it replaces.

Chapter five explores the design of actively tuned inductance cancellation systems. Active tuning is useful in applications where the parasitics may vary widely (e.g. from component to component or with operating condition.) Application to power converters is considered in which the shunt path inductance of a coupled magnetic filter can be actively controlled to form a "T-filter" with inductance cancellation. This chapter looks at trade-offs in the design of such filters that affect their performance. This chapter also offers an experimental comparison of the performances of this modified T-filter, an L-section filter, and a "zero-ripple" filter.

Chapter six extends the work in the previous five chapters by implementing inductance cancellation for multiple capacitors with a single transformer. This chapter illustrates the effective performance with test filters and through the modification of a commercial EMI filter; a filter model is also developed which can be used to analyze performance or for circuit simulations.

Chapter seven looks at additional applications of inductance cancellation. Applications considered include the improvement of sense resistors by compensating for their parasitic inductance. This serves as a good example of where inductance cancellation can be used.

Chapter eight introduces the dual cancellation approach, capacitance cancellation. In this system the parasitic capacitance of an inductor or transformer can be effectively shifted from its position in parallel to the magnetic element to a shunt path in the filter, thus improving filter performance. This technique is examined with practical elements and a limitation of this technique is identified. Two implementation methods that overcome this limitation are also proposed. This chapter concludes with a demonstration of capacitance cancellation improving the common mode performance of a commercial EMI filter.

9.2 Conclusions

This report shows that the high frequency performance of passive elements and filters can be improved by new methods that offset parasitic effects. Enhancing passive elements with inductance and capacitance cancellation is a low cost method of improving the high frequency performance of these elements. Typically, improvements of an order of magnitude or more are possible. Inductance and capacitance cancellation techniques can be used to reduce the size of filters of high switching frequency converters, making these converters more affordable.

Inductance and capacitance cancellation also provides some notable new opportunities in design. Given that these techniques will be used in a filter, the design of inductors and capacitors can be made with less regard to their parasitic inductance or capacitance. For example, toroidal inductors and transformers are typically wound with only one layer of windings in order to keep the equivalent parasitic capacitance low. Inductors or transformers can be made more volumetrically efficient by using multiple layers without regard to the additional capacitance that results, using capacitance cancellation. Likewise, some capacitors use multiple connections from the cathode or anode lead to the respective plate in order to reduce the total equivalent series inductance of the device. The manufacturing cost of the capacitor may be reduced if one does not use such connections.

Appendix A

Transformer Models

A.1   Introduction

Any time two current-carrying pathways link flux from one pathway to another a transformer is formed. Several slightly different methods of describing the transformers have been created and are commonly used. This appendix examines several different transformer models and how the parameters of these models relate to each other. Also, in Chapter 2 special "end-tapped" and "center-tapped" transformers are formed by tying two of the transformer terminals together. This three terminal transformer is converted into the "T"-model. This appendix shows the process to make that conversion.

A.2   Transformer Parameters

A transformer with two windings has three independent inductance parameters that define its operation. Several models and representations of these transformers exist and conversion from one set of parameter values to another is a straightforward process. The transformer model used in this report includes the parameters found in the terminal equations

$$\begin{bmatrix} v_1 \\ v_2 \end{bmatrix} = \begin{bmatrix} L_{11} & L_M \\ L_M & L_{22} \end{bmatrix} \frac{d}{dt} \begin{bmatrix} i_1 \\ i_2 \end{bmatrix} \tag{A.1}$$

L_{11} and L_{22} are the self inductances of the primary and secondary windings and L_M is the mutual inductance between the two windings. The standard model of the transformer [63, 64] is shown in Fig. A.1 where L_μ is the magnetizing inductance, N1:N2 is the turns ratio, and L_{l1} and L_{l2} are the leakage inductances. The conversion of these parameters to the self and mutual inductances are

$$L_{11} = L_{l1} + L_\mu \tag{A.2}$$

$$L_{22} = L_{l2} + \frac{N_2^2}{N_1^2} L_\mu \tag{A.3}$$

$$L_M = \frac{N_2}{N_1} L_\mu \tag{A.4}$$

115

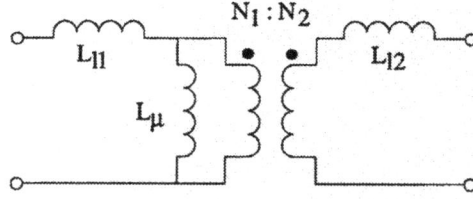

Figure A.1: The typical model of a transformer.

Another popular transformer parameters is found in PSpice [65]. This model consists of two inductances,L_p and L_s, and a coupling coefficient, k. The two self inductances, L_{11} and L_{22}, are equal to L_p and L_s respectively and the mutual inductance is

$$L_M = k\sqrt{L_{11}L_{22}} \tag{A.5}$$

The turns ratio for this transformer from the primary to secondary is the square root of L_p/L_s. The following equations can be used to convert the PSpice parameters to the ones used in Fig. A.1.

$$L_\mu = kL_p \tag{A.6}$$

$$L_{l1} = L_p(1-k) \tag{A.7}$$

$$L_{l2} = L_s(1-k) \tag{A.8}$$

A.3 Converting an End-tapped Transformer to a "T"-Model

In order to convert the standard transformer model (Fig. A.1) connected in an end-tapped manner, shown in Fig. A.2, to a "T" model we need to examine the inductive impedance across each set of terminals. If the inductance is measured across terminals A and B and terminal C is left unconnected then the inductance is simply

$$L_{AB} = L_{l2} + \frac{N_2^2}{N_1^2}L_\mu = L_{22} \tag{A.9}$$

similarly if the inductance across terminals A and C and terminal B is unconnected the inductance is

$$L_{AC} = L_{l1} + L_\mu = L_{11} \tag{A.10}$$

Finally, the inductance across terminals B and C is found while leaving terminal A open. The circuit in Fig. A.3 is the resulting circuit. The inductive impedance, L_{BC} is

$$\frac{V_t}{I_t s} = L_{BC} = L_{l2} + L_{l1} - L_\mu\left(1 - \frac{N_2}{N_1}\right)\frac{N_2}{N_1} + L_\mu\left(1 - \frac{N_2}{N_1}\right) \tag{A.11}$$

116

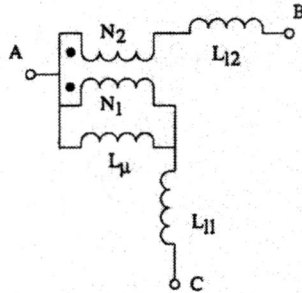

Figure A.2: A physically-based circuit model of the end-tapped coupled magnetic windings.

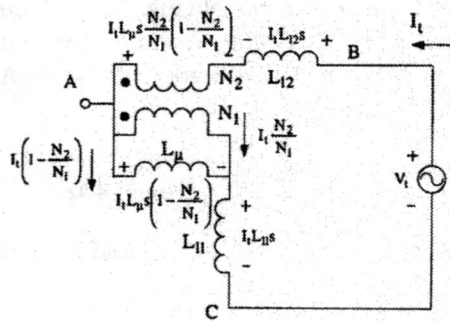

Figure A.3: The test setup to determine L_{BC}.

117

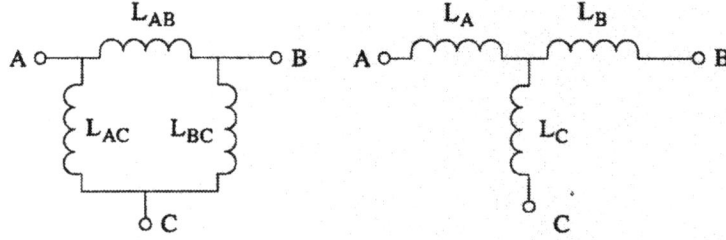

Figure A.4: The Delta and T models for a three terminal device. The inductances, L_{AB}, L_{BC}, and L_{AC} can be easily converter to the T model parameters

$$\frac{V_t}{I_t s} = L_{BC} = L_{l2} + \frac{N_2^2}{N_1^2}L_\mu + L_{l1} + L_\mu - 2L_\mu\frac{N_2}{N_1} \tag{A.12}$$

$$\frac{V_t}{I_t s} = L_{BC} = L_{AB} + L_{AC} - 2L_M \tag{A.13}$$

where

$$L_M = L_\mu\frac{N_2}{N_1} \tag{A.14}$$

These equations determine the values for the three terminal delta-model of the transformer. The final step is to convert this delta-model into a T-model as shown in Fig. A.4 The conversion results in the following values

$$L_A = L_M \tag{A.15}$$

$$L_B = L_{22} - L_M \tag{A.16}$$

$$L_C = L_{11} - L_M \tag{A.17}$$

A.4 Converting a Center-tapped Transformer to a "T"-Model

The conversion process for a center-tapped transformer to a T-model is a similar process. The standard model of a center tapped transformer is shown in Fig. A.5. Solving for the equivalent inductive impedances across the A and C terminals and the B and C terminals can be easily solved

$$L_{AC} = L_{l1} + L_\mu = L_{11} \tag{A.18}$$

$$L_{BC} = L_{l2} + \frac{N_2^2}{N_1^2}L_\mu = L_{22} \tag{A.19}$$

The final term, L_{AB}, is solved using the circuit shown in Fig. A.6. In the final inductance is

118

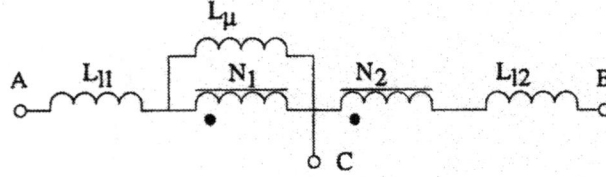

Figure A.5: A physically-based circuit model of the center-tapped coupled magnetic windings.

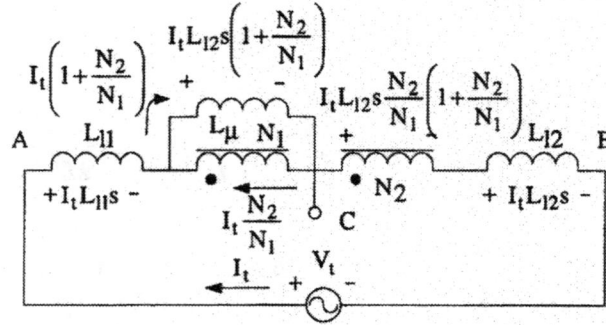

Figure A.6: The test setup to determine L_{AB}.

$$\frac{V_t}{I_t s} = L_{AB} = L_{l2} + L_{l1} + L_\mu \left(1 + \frac{N_2}{N_1}\right)\frac{N_2}{N_1} + L_\mu \left(1 + \frac{N_2}{N_1}\right) \tag{A.20}$$

$$\frac{V_t}{I_t s} = L_{BC} = L_{l2} + \frac{N_2^2}{N_1^2}L_\mu + L_{l1} + L_\mu + 2L_\mu \frac{N_2}{N_1} \tag{A.21}$$

$$\frac{V_t}{I_t s} = L_{BC} = L_{AB} + L_{AC} + 2L_M \tag{A.22}$$

Finally, these delta model terms can be converter to T-model terms resulting in

$$L_A = L_{11} + L_M \tag{A.23}$$

$$L_B = L_{22} + L_M \tag{A.24}$$

$$L_C = -L_M \tag{A.25}$$

119

Appendix B

Empirical Inductance Calculation Formulas

In this section we present empirical formulas for calculating self and mutual inductance of planar rectangular coils. In general approximate formulas can be found by curve fitting a given set of datapoints. In this case all the formulas were derived by curve fitting simulation data found using the program FastHenry. The formulas describe rectangular coils with the relative sizes and shapes that are typically needed for inductance cancellation techniques. This includes coils with dimensions between .5 and 2 inches on a side. Similar formulas can be developed for any shape or size coil.

These formulas only consider coils made up of full turns. Multiple turn coils will be considered as mutually coupled coils in series in which the total inductance LT is calculated as

$$L_T = \sum_{i=1}^{n} L_{ii} + \sum_{i=1}^{n}\sum_{j=1}^{i-1} 2L_{ij} \tag{B.1}$$

where L_{ij} is the mutual inductance between turns i and j, L_{ii} is the self inductance of turn i, and n is the number of turns.

The self-inductance of any turn can be given as

$$L_{ii} = L_{sq}K_W K_R \tag{B.2}$$

where each of these factors are defined in this appendix.

L_{sq} is the inductance of an equivalent square coil with a 100 mil trace width. The equivalent square coil is defined as a square coil with the same area as the rectangle coil. L_{sq} is defined as

$$L_{sq} = 82.25 \left[\frac{nH}{inch}\right] s_{eq}\,[inch] - 23.51\,[nH] \tag{B.3}$$

$$s_{eq} = \sqrt{LW} \tag{B.4}$$

where s_{eq}, L, W are the length of the side of the equivalent square coil, the length of the rectangular coil, and the width of the rectangular coil. This formula applies to square coils with areas between .25 and 4 square inches.

The factor K_W is used to compensate for the width, w, of the trace. This factor depends on the width of the trace and the length of a side of the equivalent square coil. K_w is defined as

$$K_W = B + A \ln(w[mil]) \tag{B.5}$$

where

$$A = 0.0833 \ln(s_{eq}[in]) - 0.3297 \tag{B.6}$$

$$B = 1 - A \ln(100) \tag{B.7}$$

For rectangular coils the factor K_R is needed. This factor depends on the ratio of the sides of the rectangle k_r. (Note that k_r will always be greater than 1.)

$$K_R = 1 + (k_r - 1)0.055 \tag{B.8}$$

The mutual inductance between two coils is given by

$$L_{ij} = L_M K_{R-ave} K_z \tag{B.9}$$

L_M is the mutual inductance assuming both coils are square and that they are on the same layer. K_{R-ave} and K_Z modify this number to compensate if the coils have unequal sides or if the coils are on different layers of the PCB. L_M is given by

$$L_M = C_M e^{D_M s_{eq2}} \tag{B.10}$$

$$C_M = E_M \ln(s_{eq1}) + F_M \tag{B.11}$$

$$D_M = G_M \ln(s_{eq1}) + H_M \tag{B.12}$$

$$E_M = 0.0092(w_1) - 0.0008 \tag{B.13}$$

$$F_M = -0.005(w_1) + 1.4426 \tag{B.14}$$

$$G_M = -0.0096(w_1) - 1.8523 \tag{B.15}$$

$$H_M = 0.006(w_1) + 3.5207 \tag{B.16}$$

where s_{eq1} is the length of the side of the equivalent square coils in inches for the larger turn, s_{eq2} is the equivalent length for the smaller turn, and w_1 is the trace width of the first coil in mils. If the 2 coils have different trace width then define the coil with the smaller width as coil 1. L_M will be in nH.

K_{R-ave} is the average of the rectangular coil constant K_R and is given as

$$K_{R-ave} = (K_R|_{coil1} + K_R|_{coil2})\frac{1}{2} \tag{B.17}$$

121

The last factor, K_Z, accounts for displacement between coils on different layers. Typical board spacing is either 31 or 62 mil spacing. This factor is approximated as a constant factor for a given spacing K_Z is 0.99 for 31 mil spacing and 0.975 for 62 mil spacing.

With these inductance formula the following case was examined. A two-turn inductor has a trace width of 150 mils and the first turn has sides of lengths 1200 mils and 1000 mils, the second turn has sides of 800 mils and 600 mils.

For the first turn the terms to calculate the inductance are:

$s_{eq1} = 1.095$ inches $\quad L_{eq} = 66.6$ nH

$\quad A = -0.3220 \qquad B = 2.4829$

$\quad K_W = 0.8694 \qquad k_r = 1.2$

$\quad K_R = 1.011 \qquad L_{11} = 58.5$ nH

For the second turn the terms to calculate the inductance are:

$s_{eq2} = 0.6928$ inches $\quad L_{eq} = 33.47$ nH

$\quad A = -0.3603 \qquad B = 2.659$

$\quad K_W = 0.8536 \qquad k_r = 1.333$

$\quad K_R = 1.0183 \qquad L_{22} = 29.09$ nH

The terms to get the mutual inductance are:

$\quad E_m = 1.3792 \qquad F_m = 0.6926$

$\quad G_m = -3.2923 \qquad E_m = -4.4207$

$\quad C_m = 0.81776 \qquad D_m = 4.1219$

$\quad L_m = 14.22 \qquad K_{R-ave} = 1.01465$

$\quad K_Z = 1 \qquad L_{12} = 14.43$

The inductance for the inductor becomes 116.45 nH. The FastHenry prediction for this inductor is 123.37 nH.

The data used to derive these equations consisted mainly of simulations using two 1-turn windings. These empirical equations do not take into consideration the effects of nearby traces, or other nearby turns. When calculating the inductances for a transformer with many turns these equations (and other empirical and analytical equations) will calculate the inductance of 2 turns of the transformer at a time and appropriately combine these values. Each calculation will assume no other turns exist and this assumption will lead to a overestimation of the final inductance values. These equations adequately predict the inductances for cases with low number of turns but computer simulations should be used whenever an accurate estimation of the inductances associated with an air-core transformer with about 8 turns.

Appendix C

Cost Estimation for Integrated Filter Elements

The following table lists the expected costs of processes and materials needed to construct integrated filter elements as described in Section 4.4. All the numbers were generated as estimates with the help of Professor Rich Roth from the Technology and Policy Program at M.I.T. unless otherwise stated. These numbers were generated to get a rough estimate of the cost of mass producing integrated filter elements and are not actual quotes. The process assumes that the windings are stamped out in copper, a layer of insulation is attached to one side of the transformer before it is stamped and folded into the final shape, that a potting compound is used, and that a new welding head in needed to make the interconnection.

The pricing is divided into nine sections, one table for constants used by the other sections, five sections based on the cost of the machines needed for the production line, and three sections based on the material used in the construction. It is assumed that the production line will be mass producing the integrated filter elements used in Section 4.4. It is also assumed that the production line will take 6 by 12 inch copper sheets as the input stock and that the winding pattern will remain the same as the prototype. Because the same transformer is used then the same amount of potting compound is needed. The winding pattern, amount of potting material needed, and choice of stock to cut the winding pattern from are all manufacturing questions that will affect the final cost of the integrated filter element. These particular design choices are used to get a very rough estimate of the cost.

The first table lists constants. We are assuming that the factory will have two eight hour shifts and will be operating 250 days a years. There is a time efficiency which is the percentage of working hours that the line is running. Also there are a number of winding patterns that can be formed from one 6 by 12 inch sheet of copper.

The first process is the creation of a winding pattern. From one sheet of copper 63 winding patterns can be stamped out at one time. The stamping machine is assumed to operate once every three seconds. We estimated the cost of the machine to be 500 thousand dollars and the machine will have a 10 year lifetime. With this information we found the cost per unit of the machine. Results are in Table C.2.

The second process is applying an insulation layer to one side of the transformer winding. It will be assumed that this layer is applied before the copper sheet is stamped. The sheet is sprayed with adhesive and a layer of insulation foil is applied. The insulation foil can be the same material

Item Number	Item	Value
A_1	Units Per Sheet	63
A_2	Work Hours Per Day	16
A_3	Work Days per Year	250
A_4	Time Efficiency	0.85

Table C.1: Constants

Item Number	Item	Value	Formula
B_1	Stamping Machine	500,000	
B_2	Machine Life Time (years)	10	
B_3	Stamps Per Minute	20	
B_4	Stamps Per Year	4080000	$60 A_2 A_3 A_4 B_3$
B_5	Cost of Machine Per Unit	0.000194522	$\frac{B_1}{A_1 B_2 B_4}$

Table C.2: Stamping machine cost

used in film-foil or metallized film capacitors. Since this material is readily available at a capacitor factory, and so little is needed per filter element, the cost of foil is ignored. The machine used for spraying adhesive was estimated to cost 50 thousand dollars and to have a lifetime of 10 years. This machine will produce the same number of units as the previous step. The cost per unit is shown in Table C.3

The third step in the process is to fold the winding pattern into a transformer. The folding machine is assumed to cost 50 thousand dollars and have a lifetime of 10 years. It is assumed that the machine can make a fold every 3 seconds. The cost per unit for this machine is shown in Table C.4.

The forth step will be to inject potting material into a mold. The machine was assumed to have space for 20 molds and 1 minute is needed for each mold to form, thus the machine can process 20 units a minute. The machine is assumed to cost 50 thousand dollars and to have a lifetime of 10 years. The per unit cost is shown in Table C.5.

The last step is to wield the interconnections. It will be assumed that the increase from 2 interconnection to 4 will only result in the welding machine having a new head that can make four connections with each operation. This new head for the welding machine is assumed to cost 10 thousand dollars and to have a lifetime of 10 years. The welding machine is assumed to make

Item Number	Item	Value	Formula
C_1	Spraying Machine	50000	
C_2	Machine Life Time (years)	10	
C_3	Units Per Year	257040000	$A_1 B_4$
C_4	Cost Per Unit	0.0000194522	$\frac{C_1}{C_2 C_3}$

Table C.3: Spraying machine cost

Item Number	Item	Value	Formula
D_1	Folding Machine	50000	
D_2	Machine Life Time (years)	10	
D_3	Folds Per Minute	20	
D_4	Folds Per Year	4080000	$60A_2A_3A_4D_3$
D_5	Cost Per Unit	0.00122549	$\frac{D_1}{D_2D_4}$

Table C.4: Folding machine cost

Item Number	Item	Value	Formula
E_1	Potting Material Injector Machine	50000	
E_2	Machine Life Time (years)	10	
E_3	Molds Per Minute	20	
E_4	Molds Per Year	4080000	$60A_2A_3A_4E_3$
E_5	Cost Per Unit	0.00122549	$\frac{E_1}{E_2E_4}$

Table C.5: Potting material injector machine cost

connections at a rate of once every two seconds. The cost per unit is in Table C.6.

Table C.7 shows the expected price for the copper used. The price of copper ingot is taken from the London Metal Exchange (on 4/15/2004) and the weight of a sheet of a sheet of copper and a winding pattern is measured. The amount of copper left over from the stamping process will be sold off as scrap. It is assumed that the price of scrap copper is 20% of the cost of the ingot. The total cost is then used to find the cost per unit. Note that this estimate is based on a 10 mil thick winding, with an equivalent dc current rating of 2.58 amperes assuming a $500A/cm^2$ current density. The copper cost scales linearly with dc current rating.

Adhesive is needed to connect the insulator to the copper sheets. This adhesive can be an industrial grade aerosol adhesive that is used in the spraying machines. The spray will have to cover the whole area of the copper sheet and we assumed the thickness will be 0.1 mils. An estimate of the cost of the adhesive is based upon a quote (for small quantities) on a 3M industrial adhesive (Super 77 Spray Adhesive). The cost of the adhesive per unit is shown in Table C.8.

Potting material will be used to encapsulate the transformer and to set all the spacing distances. The pricing of the potting material is from a website (for 60 gallons) on a West system potting

Item Number	Item	Value	Formula
F_1	Welding Machine Head	10000	
F_2	Machine Life Time (years)	10	
F_3	Welds Per Minute	30	
F_4	Welds Per Year	6120000	$60A_2A_3A_4F_3$
F_5	Cost Per Unit	0.000163399	$\frac{F_1}{F_2F_4}$

Table C.6: Welding machine head cost

Item Number	Item	Value	Formula
G_1	Copper Price Per Tonne	2932.5	
G_2	Weight of Sheet (kg)	0.1043	
G_3	Price Per Sheet	0.30585975	$\frac{G_1 G_2}{1000}$
G_4	Weight Per Unit (kg)	0.0006	
G_5	Scrap copper per sheet (kg)	0.0665	$G_2 - A_1 G_4$
G_6	Scrap Resale Rate	0.2	
G_7	Copper Scrap Sell Price per Tonne	586.5	$G_1 G_6$
G_8	Scrap Price Per Sheet	0.03900225	$\frac{G_7 G_5}{1000}$
G_9	Net Price Per Sheet	0.2668575	$G_3 - G_8$
G_{10}	Price for Copper Per Unit	0.004235833	$\frac{G_9}{A_1}$

Table C.7: Copper cost for an integrated filter with a dc current rating of 2.58 A.

Item Number	Item	Value	Formula
H_1	Spray Price Per cm^2	0.0000504925	
H_2	Area of Sheet (cm^2)	464.5152	
H_4	Cost Per Sheet	0.023454545	$\frac{H_1 H_2}{10000}$
H_5	Cost Per Unit	0.000372294	$\frac{H_4}{A_1}$

Table C.8: Adhesive cost

resin (105-E and 205-E hardener). Note that other types of potting material should have lower costs. The volume of the potting material is taken from measurements of the prototype. The price per unit is given in Table C.9.

The machine costs and the materials costs are summarized in Table C.10. These figures results in a rough estimate of the price of upgrading a capacitor to an integrated filter element with a dc current rating of 2.58 A to be 0.02395. An incremental cost for an integrated device having a 3A dc current rating would be 0.02464 or 0.000688 more than the previous case.

Item Number	Item	Value	Formula
I_1	Potting Material Price Per Liter	11.66	
I_2	Potting Area (cm^2)	6.193536	
I_3	Potting Height (cm)	0.2286	
I_4	Potting Volume (cm^3)	1.41584233	$I_2 I_3$
I_5	Cost Per Unit	0.016515705	

Table C.9: Potting material cost

126

Item Number	Item	Value
	Machine Costs Per Unit	
J_1	Stamping Machine	0.000194522
J_2	Spraying Machine	0.0000194522
J_3	Folding Machine	0.00122549
J_4	Potting Material Injector Machine	0.00122549
J_5	Welding Machine Head Upgrade	0.000163399
J_6	Total Machine Cost	0.002828354
	Material Costs Per Unit	
J_7	Copper	0.004235833
J_8	Adhesive	0.000372294
J_9	Potting Material	0.016515705
J_{10}	Total Materials Cost	0.021123832
J_{11}	Total Cost	0.023952186

Table C.10: Adhesive cost

Appendix D

Three-Port Tapped-Inductor Extended Cantilever Model Transfer Function

Equation (D.1) gives the analytic solution of the transfer function from input voltage V_{in} to output voltage V_l for the circuit in Fig. 6.19. The result was found using direct circuit analysis, with the source network consisting of an input voltage source V_{in} with series impedance z_s, and a load network comprised of an impedance z_l. T_2 and T_3 are arbitrary impedances representing the two capacitors.

$$H(s) = \frac{\begin{aligned}(T_2T_3 - Z_{13}T_3 - Z_{12}T_3 - Z_{23}T_2 \\ - Z_{13}T_2 + Z_{12}Z_{23} - Z_{13}Z_{22})\,z_l\end{aligned}}{\begin{aligned} &T_3z_lz_s + T_2z_lz_s + Z_{22}z_lz_s + T_2T_3z_s \\ &+ Z_{33}T_3z_s + 2Z_{23}T_3z_s + Z_{22}T_3z_s \\ &+ Z_{33}T_2z_s + Z_{22}Z_{33}z_s - Z_{23}^2z_s \\ &+ T_2T_3z_l + Z_{11}T_3z_l + Z_{22}T_2z_l + 2Z_{12}T_2z_l \\ &+ Z_{11}T_2z_l + Z_{11}Z_{22}z_l - Z_{12}^2z_l \\ &+ Z_{33}T_2T_3 + 2Z_{23}T_2T_3 + Z_{22}T_2T_3 \\ &+ 2Z_{13}T_2T_3 + 2Z_{12}T_2T_3 + Z_{11}T_2T_3 \\ &+ Z_{11}Z_{33}T_3 + 2Z_{11}Z_{23}T_3 + Z_{11}Z_{22}T_3 \\ &- Z_{13}^2T_3 - 2Z_{12}Z_{13}T_3 - Z_{12}^2T_3 + \\ &Z_{22}Z_{33}T_2 + 2Z_{12}Z_{33}T_2 + Z_{11}Z_{33}T_2 \\ &- Z_{23}^2T_2 - 2Z_{13}Z_{23}T_2 - Z_{13}^2T_2 \\ &+ Z_{11}Z_{22}Z_{33} - Z_{12}^2Z_{33} - Z_{11}Z_{23}^2 \\ &+ 2Z_{12}Z_{13}Z_{23} - Z_{13}^2Z_{22} \end{aligned}} \qquad (D.1)$$

Bibliography

[1] P.T. Krein. *Elements of Power Electronics*. Oxford university Press, New York, 1998. pages 381-392.

[2] T.K. Phelps and W.S. Tate. "Optimizing Passive Input Filter Design". In *Proc. of the 6th National Solid-State Power Conversion Conf.*, pages G1-1 – G1-10, May 1979.

[3] M.J. Nave. *Power Line Filter Design for Switched-Mode Power Supplies*. Van Nostrand Reinhold, New York, 1991.

[4] G.B. Crouse. "Electrical Filter". U.S. Patent No. 1,920,948, Aug. 1 1933.

[5] D.C. Hamill and P.T. Krein. "A 'Zero' Ripple Technique Applicable to Any DC Converter". In *1999 IEEE Power Electronics Specialists Conference*, pages 1165–1171, June 1999.

[6] S. Feng, W.A. Sander, and T.G. Wilson. "Small-Capacitance Nondissipative Ripple Filters for DC Supplies". In *IEEE Trans. Mag.*, volume 6, pages 137–142, March 1940.

[7] R.P. Severns and G.E. Bloom. "*Modern DC-to-DC Switchmode Power Converter Circuit*". Van Nostrand Reinhold, New York, 1985.

[8] J.W. Kolar, H. Sree, N. Mohan, and F.C. Zach. "Novel Aspecs of an Application of 'Zero'-Ripple Techniques to Basic Converte Topologies". In *1997 IEEE Power Electronics Specialists Conference*, pages 796–803, June 1997.

[9] G.E. Bloom and R. Severns. "The Generalized Use of Integrated Magnetics and Zero-Ripple Techniques in Switchmode Power Converters". In *1984 IEEE Power Electronics Specialists Conference*, pages 15–33, June 1984.

[10] S. Senini and P.J. Wolfs. "The Coupled Inductor Filter: Analysis and Design for AC Systems". In *IEEE Trans. Industrial Electronics*, volume 45, pages 574–578, Aug. 1998.

[11] J.W. Phinney. "Filters with Active Tuning for Power Applications". S.m., Massachusetts Institute of Technology, May 2001. test.

[12] J.W. Phinney and D.J. Perreault. "Filters with Active Tuning for Power Applications". In *2001 IEEE Power Electronics Specialists Conference*, pages 363–370, June 2001.

[13] D.L. Logue and P.T. Krein. "Optimization of Power Electronic Systems Using Ripple Correlation Control: A Dynamic Programming Approach". In *2001 IEEE Power Electronics Specialists Conference*, pages 459–464, June 2001.

[14] R. Heartz and H. Buelteman. "The Application of Perpendicularly Superposed Magnet Fields". In *AIEE Transactions, Part I*, volume 74, pages 655–660, Nov. 1955.

[15] N. Saleh. "Variable Microelectronic Inductors". In *IEEE Trans. Comp. Hybrid, and Mfg. Tech.*, volume CHMT-1, pages 118–124, March 1978.

[16] H.J. McCreary. "The Magnetic Cross Valve". In *AIEE Transactions, Part II*, volume 70, pages 1868–1875, 1951.

[17] F.J. Beck and J.M. Kelly. "Magnetization in Perpendicularly Superposed Direct and Alternating Fields". In *Journal of Applied Physics*, volume 19, pages 551–562, June 1948.

[18] A. Kamon, L.M. Silveira, C. Smithhisler, and J. White. *FastHenry USER'S GUIDE*. MIT, Cambridge, 1996.

[19] R. Reeves. "Choke-Capacitor Hybrid as a Flourescent Lamp Ballast". In *Proc. IEE*, volume 122, pages 1151–1152, Oct. 1975.

[20] R. Reeves. "Inductor-Capacitor Hybrid". In *Proc. IEE*, volume 122, pages 1323–1326, Nov. 1975.

[21] P.N. Murgatroyd and N.J. Walker. "Lumped-Circuit Model for Inductor-Capacitor Hybrid". In *Electronic Letters*, volume 12, pages 2–3, 1976 Jan.

[22] R.J. Kemp, P.N Murgatroyd, and N.J. Walker. "Self Resonance in Foil Inductors". In *Electronic Letters*, volume 11, pages 337–338, July 1975.

[23] M. Ehsani, O.H. Stielau, J.D. (Van Wyk), and I.J. Pitel. "Integrated Reactive Components on Power Electronics Circuits". In *IEEE Trans. Power Electronics*, volume 8, pages 205–215, April 1993.

[24] M. Ehsani, P. Le Polles, M.S Arefeen, I.J. Pitel, and J.D. Van Wyk. "Computer-Aided Design and Application of Integrated LC Filters". In *IEEE Trans. Power Electronics*, volume 11, pages 182–190, Jan. 1996.

[25] I.W. Hofsajer, J.A. Ferreira, and J.D. van Wyk. "Design and Analysis of Planer Integrated L-C-T Components for Converters". In *IEEE Trans. Power Electronics*, volume 15, pages 1221–1227, Nov. 2000.

[26] F. Wilmot, E. Labouré, F. Costa, C. Joubert S. Faucher, and F. Forest. "Design Optimization and Electromagnetic Modeling of Integrated Passive Components for Power Electronic". In *2001 IEEE Power Electronics Specialists Conference*, pages 1932–1937, June 2001.

[27] L. Zhao, J.T. Strydom, and J.D. van Wyk. "An Integrated Resonent Module for a High-Power Soft-Switching Converter". In *2001 IEEE Power Electronics Specialists Conference*, pages 1944–1948, June 2001.

[28] K. Laouamri, J.C. Crebier, J.-P. Ferrieux, and T. Chevalier. "Construction and Modeling of Integrated LCT Structure for PFC Resonant Converter". In *2001 IEEE Power Electronics Specialists Conference*, pages 1949–1954, June 2001.

[29] J.C. Crebier, T. Chevalier, K. Laouamri, and J.-P. Ferrieux. "Study and Analysis of Wounded Integrated L-C Passive Components". In *2001 IEEE Power Electronics Specialists Conference*, pages 2137–2142, June 2001.

[30] C.M. Zierhofer. "Geometric Approach for Coupling Enhancement of Magnetic Coupled Coils". In *IEEE Trans. On Biomedical Engineering*, volume 43, pages 708–714, July 1996.

[31] W.G. Hurley and M.C. Duffy. "Calculations of Self and Mutual Impedances in Planar Magnetic Structures". In *IEEE Trans. on Magnetics*, volume 31, pages 2416–2422, July 1995.

[32] F.W. Grover. *Inductance Calculations: Working Formula and Tables*. Dover Publishing, Inc, New York, 1946.

[33] H.E. Bryan. "Printed inductors and capacitors". In *Tele-Tech and Electronic Industries*, volume 14, page 68, Dec. 1955.

[34] H.G. Dill. "Designing inductors for thin-film application?". In *Electronic Design*, pages 52–59, Feb. 1964.

[35] H.A. Wheeler. "Simple inductance Formulas for Radio Coils". In *Procedings of the I.R.E.*, volume 16, October 1928.

[36] C.R. Sullivan and A.M. Kern. "Capacitors with Fast Current Switching Require Distributed Models ". In *2001 IEEE Power Electronics Specialists Conference*, pages 1497–1503, June 2001.

[37] Henry W. Ott. *Noise Reduction Techniques in Electronic Systems*. John Wiley & Sons, 2nd edition, 1988.

[38] T.C. Neugebauer, J.W. Phinney, and D.J. Perreault. Filters and components with inductance cancellation. 40(2):483–491, March-April 2004.

[39] A. Kamon, L.M. Silveira, C. Smithhisler, and J. White. *FastHenry USER'S GUIDE*. MIT Research Laboratory of Electronics, Cambridge, MA 02139 U.S.A., 3.0 edition, November 1996.

[40] T.C. Neugebauer and D.J. Perreault. Filters with inductance cancellation using printed circuit board transformers. 19(3):591–602, May 2004.

[41] S. Wang, F.C. Lee, D.Y. Chen, and W.G. Odendaal. Effects of parasitic parameters on emi filter performance. 19(3):869–877, May 2004.

[42] Tyco Electronics, Libertyville, Illinois 60048 U.S.A. *CORCOM Product Guide Catalog 1654001*, March 2004.

[43] D. Maksimovic, R.W. Erickson, and C. Griesbach. Modeling of cross-regulation in converters containing coupled inductors. 15(4):607–615, July 2000.

[44] K. Changtong, R.W. Erickson, and D. Maksimovic. A comparison of the ladder and full-order magnetic models. In *Proceedings of the IEEE Power Electronics Specialists Conference*, volume 4, pages 2067–2071, Vancouver, BC, June 2001.

[45] J.G. Hayes, N. O'Donovan, and M.G. Egan. The extended t model of the multiwinding transformer. In *Proceedings of the IEEE Power Electronics Specialists Conference*, volume 3, pages 1812–1817, June 2004.

[46] K.D.T. Ngo, S. Srinivas, and P. Nakmahachalasint. Broadband extended cantilever model for magnetic component windings. 16(4):551–557, July 2001.

[47] K.D.T. Ngo and A. Gangupomu. Improved method to extract the short-circuit parameters of the becm. *IEEE Power Electronics Letters*, 1(1):17–18, 2003.

[48] H.W. Ott. *Noise Reduction Techniques in Electronic Systems*. John Wiley and Sons, New York, 1998. Chapter 5.

[49] A. Massarini, M.K. Kazimierczuk, and G. Grandi. Lumped Parameter Models for Single- and Multiple-Layer Inductors. In *1996 IEEE Power Electronics Specialists Conference*, pages 295–301, June 1996.

[50] A. Massarini and M.K. Kazimierczuk. Self-Capacitance of Inductors. In *IEEE Transactions on Power Electronics*, volume 12, pages 671–676, July 1997.

[51] M.K. Kazimierczuk, G. Sancineto, G. Grandi, U. Reggiani, and A. Massarini. High-Frequency Small-Signal Model of Ferrite Core Inductors. In *IEEE Transactions on Magnetics*, volume 35, pages 4185–4191, Sept. 1999.

[52] W. Xin, F.N.K. Poon, C.M. Lee, M.H. Pong, and Z. Qian. A Study of Common Mode Noise in Switching Power Supply from a Current Balancing Viewpoint. In *IEEE International Conference on Power Electronics and Drive Systems*, pages 621–625, 1999.

[53] W. Xin, M.H. Pong, Z.Y. Lu, and Z.M. Qian. Novel Boost PFC with Low Common Mode EMI: Modeling and Design. In *2000 IEEE Applied Power Electronics Conference*, pages 178–181, 2000.

[54] M. Kchikach, Z.M. Qian, X. Wu, and M.H. Pong. The Influences of Parasitic Capacitances on the Effectiveness of Anti-Phase Technique for Common Mode Noise Suppression. In *IEEE International Conference on Power Electronics and Drive Systems*, volume 1, pages 115–120, Oct. 2001.

[55] D. Cochrane, D.Y. Chen, and D. Boroyevic. Passive Cancellation of Common-Mode Noise in Power Electronic Circuits. In *IEEE Transactions on Power Electronics*, volume 18, pages 756–763, May 2003.

[56] M. Shoyama, T. Okunaga, G. Li, and T. Ninomiya. Balanced Switching Converter to Reduce Common-Mode Noise. In *2001 IEEE Power Electronics Specialists Conference*, pages 451–456, June 2001.

[57] M. Shoyama, M. Ohba, and T. Ninomiya. Balanced Buck-Boost Switching Converter to Reduce Common-Mode Conducted Noise. In *2002 IEEE Power Electronics Specialists Conference*, pages 2056–2061, June 2002.

[58] T.C. Neugebauer, J.W. Phinney, and D.J. Perreault. Filters and Components with Inductance Cancellation. In *2002 IEEE Industry Applications Society Annual Meeting*, pages 939–947, 2002.

[59] T.C. Neugebauer and D.J. Perreault. Filters and Components using Printed Circuit Board Transformers. In *2003 IEEE Power Electronics Specialists Conference*, pages 272–282, 2003.

[60] F.E. Terman. *Electronic and Radio Engineering*. McGraw-Hill Book Co., New York, fourth edition, 1955.

[61] T.H. Lee. *The Design of CMOS Radio-Frequency Integrated Circuits*. Cambridge University Press, New York, 1998. Chapter 1: A Nonlinear History of Radio, pp. 16-17.

[62] A.C. Chow and D.J. Perreault. Design and Evaluation of a Hybrid Passive/Active Ripple Filter with Voltage Injection. In *IEEE Transactions on Aerospace and Electronic Systems*, volume 39, pages 471–480, April 2003.

[63] J.G. Kassakian, M.F. Schlecht, and G.C. Verghese. *Principles of Power Electronics*. Addison-Wesley, Reading, Massachusetts, 1991. page 592.

[64] R.W. Erickson and D. Maksimović. *Fundamentals of Power Electronics*. Kluwer Academic Publishers, Norwell, Massachusetts, Second edition, 2001. page 505.

[65] M.H. Rashid. *SPICE for Circuit and Electronics using PSpice*. Prentice Hall, Englewood Cliffs, New Jersey, Second edition, 1995. page 106.

www.ingramcontent.com/pod-product-compliance
Lightning Source LLC
Chambersburg PA
CBHW080559220326

41599CB00032B/6541